Altered genes

Altered genes

Reconstructing nature: the debate

edited by

Richard Hindmarsh
Geoffrey Lawrence
Janet Norton

ALLEN & UNWIN

First published in 1998 by
Allen & Unwin
9 Atchison Street
St Leonards NSW 1590
Australia
Phone: (61 2) 8425 0100
Fax: (61 2) 9906 2218
E-mail: frontdesk@allen-unwin.com.au
Web: http://www.allen-unwin.com.au

National Library of Australia
Cataloguing-in-Publication entry:

Altered genes: reconstructing nature.

Bibliography.
Includes index.
ISBN 1 86448 795 X.

1. Genetic engineering—Environmental aspects—Australia. 2. Genetic engineering—Social aspects—Australia. 3. Genetic engineering—Political aspects—Australia. 4. Genetic engineering—Australia—Moral and ethical. I. Hindmarsh, R. A. II. Lawrence, G. A. (Geoffrey A.). III. Norton, Janet, 1952– .

174.957

Set in 9.5/12 pt Arrus BT by DOCUPRO, Sydney
Printed by Ligare Pty Ltd, Sydney

10 9 8 7 6 5 4 3 2 1

Contents

Tables and figures

TABLES

FIGURES

Contributors

Jean Christie is Director of International Liaison for RAFI (the Rural Advancement Foundation International), an international non-governmental organisation based in Ottawa, Canada. For twenty years RAFI has done research and policy advocacy on issues relating to biodiversity, biotechnology, intellectual property rights, and indigenous peoples' knowledge. With a BA in Political Science from Queens University (Canada) Jean was—prior to joining RAFI—Executive Director of Inter Pares, a Canadian non-profit organisation working in international cooperation for development.

Stephen Crook is Professor of Sociology at James Cook University of North Queensland, with interests in the sociology of culture, social theory and the sociology of change. He is the author of *Modernist Radicalism and its Aftermath* (1991), co-author of *Postmodernization: Change in Advanced Society* (1992) and editor of *Adorno: the Stars Down to Earth* (1994). Stephen is also a contributor to, and co-editor of, *Ebbing of the Green Tide?: Environmentalism, Public Opinion and the Media in Australia* (forthcoming). He is currently researching the sociocultural regimes through which technological and other risks are managed, and the relation of those regimes to more general patterns of order and change.

Astrid Gesche is based in the School of Humanities, Queensland University of Technology, where she recently completed a QUT Postdoctoral Fellowship as a bio-ethicist. Astrid holds a science degree in immunology and cell biology. While she supports the molecular genetic revolution that has entered the medical arena, Astrid considers it her responsibility to explore and debate the ethical issues that arise as the field develops. She is a member of the New York Academy of Sciences, the American Association for the Advancement of Science, the World Futures Federation, and the Australian Bioethics Association.

Richard Hindmarsh is an environmental researcher from the social sciences. Based at Griffith University's School of Australian Environmental Studies, Brisbane, he is the recipient of an Australian Research

Council Postdoctoral Fellowship. For over a decade he has specialised in researching and writing (with over fifty academic and lay publications) on genetic engineering issues, and is a well-known commentator in the debate. Currently, Richard is continuing his research into historical, philosophical, social, ethical, and policy aspects of genetic engineering—as well as into the alternatives to genetic engineering—and is developing holistic technology assessment modelling in the context of sustainable futures.

Geoffrey Lawrence is Foundation Professor of Sociology, and Executive Director of the Institute for Sustainable Regional Development, at Central Queensland University. He has written widely about rural social issues in Australia and has conducted many national research projects, including an ARC-funded study of agro-biotechnologies. Books include: *Capitalism and the Countryside* (1987), *Agriculture, Environment and Society* (1992), *The Environmental Imperative* (1995), *Social Change in Rural Australia* (1996), and *Globalisation and Agri-food Restructuring* (1996).

Rosaleen Love teaches professional writing in the Department of Communication and Language Studies at the Victoria University of Technology. She is the editor of *If Atoms Could Talk* (1987), and the author of *The Total Devotion Machine* (1989) and *Evolution Annie* (1993). Her most recent essays include *Bubbles in the Cosmic Saucepan* and *Fantasy and Communication Futures* (forthcoming).

Janet Norton is a doctoral candidate in sociology in the Faculty of Arts, Health and Sciences at Central Queensland University. She is attached to the University's Rural Social and Economic Research Centre, where she is undertaking a national study of Australian consumers' attitudes to genetically-engineered foods and the changing role of science in society. Janet is the co-author of several articles on the sociology of food biotechnologies in Australia.

Bob Phelps is Director of the Australian GeneEthics Network—based in Melbourne. He is a leading participant in Australian and overseas policy forums on biosafety, food labelling standards and sustainable systems and has taken part in several conferences and inquiries including the Human Genome and Human Genome Diversity Projects. He promotes public education and discussion of the environmental, social and ethical impacts of gene technologies and their products. Bob is registered with the Federal Department of Industry and Science as a Science and Technology Communicator.

Robyn Rowland, Associate Professor in Women's Studies at Deakin University Victoria, has researched the area of reproductive technology over fifteen years, and has addressed meetings of MPs and State government committees in Australia and overseas. She is the author of the book *Living Laboratories: Women and Reproductive Technologies* (1992) and has been the Australian editor of *Issues in Reproductive and Genetic Engineering: Journal of International Feminist Analysis* and *Women's Studies International Forum*. In 1996, Robyn received an Order of Australia for distinguished service in the areas of women's health and higher education.

Anna Salleh is Principal Researcher in food, health and environment for the Australian Consumers' Association. She has worked in print media, radio and television, with a special interest in science and public policy. Recently she researched a documentary on transgenic crops for ABC TV's *Quantum* programme. Anna holds a BSc from Sydney University and a Master of Arts (Journalism) from Sydney's University of Technology, and teaches specialist reporting in science, medicine and environment.

David Suzuki is an award winning scientist, environmentalist and broadcaster. Currently a Professor at the University of British Columbia, he is familiar to the public as the host of CBC's *Planet for the Taking* and *The Nature of Things*. He is the author of twenty-eight books including *Introduction to Genetic Analysis*. David is recognised as a world leader in sustainable ecology and is a recipient of Unesco's Kalinga Prize for Science, the United Nations Environment Medal, and the Global 500. He is a member of the Royal Society of Canada and the Order of Canada.

Nicholas Tonti-Filippini is a doctoral candidate at Melbourne University where he is researching 'The Concept of Dignity in the International Human Rights Instruments'. He was a consultant to Unesco's International Bioethics Committee on the Draft Declaration on the Human Genome and Human Rights, and participated in the Paris meeting which launched the Declaration. He is a member of the Australian Bioethics Association, the International Bioethics Association, the International Ethics Association and the Moral Theologians Association of Australia and New Zealand.

Frank Vanclay is Senior Lecturer in Sociology at Charles Sturt University (Wagga Wagga, New South Wales), Associate Director of the Centre for Rural Social Research, and a Project Leader with the Cooperative Research Centre for Viticulture. He has conducted studies on environmental management and impact assessment and has interests in research methods and health policy. Books include: *With a Rural Focus* (1995), *The Environmental Imperative* (1995), *Environmental and Social Impact Assessment* (1995) and *Critical Landcare* (1997).

Tiffany White holds a Masters degree in Communication, and manages public relations for Nepean Hospital, Western Sydney. Her interest in biotechnology developed from studying—in a critical manner—the tone and frequency of science-based news stories, and the social implications of recombinant-DNA technology in the wider literature. She became concerned that both sides of the story about biotechnology were not being told, and concluded that the public was not developing a balanced understanding of biotechnology.

Peter R. Wills is a theoretical biologist and Associate Professor in the Department of Physics at the University of Auckland, New Zealand. His research interests include the molecular biological aetiology of spongiform encephalapothies, the thermodynamic basis of biochemical processes, and the physical origin of evolutionary self-organisation. He is best-known to the general public as a political activist who has been critical of military applications of technology, especially in the service of nuclear, biological and chemical warfare. He has also engaged in academic work aimed at eliciting general acknowledgment of the legitimacy of non-scientific epistemologies, especially from indigenous Pacific cultures.

Acknowledgments

We would like to express our appreciation to Janét Moyle amd Emma Cotter for providing valuable feedback on the manuscript, and to Sandra Henderson, Kees Hulsman and Dimity Lawrence for their comments on particular chapters. An Australian Research Council Post-doctoral Fellowship held by Richard Hindmarsh during the course of the book's preparation—and support from the host facilities of the School of Australian Environmental Studies, Griffith University and from Central Queensland University—was greatly appreciated. The research described in chapter 13 was undertaken using funds provided by Central Queensland University in the form of two University Research Grants and a University Postgraduate Research Award. The chapter by Robyn Rowland is an updated and amended version of chapter 2 of her book *Living Laboratories: Women and Reproductive Technologies*.

Richard Hindmarsh
Geoffrey Lawrence
Janet Norton

Introduction: A geneticist's reflections on the new genetics

DAVID SUZUKI

I am delighted and honoured to have the opportunity to write an Introduction for this book. Apart from having adopted Australia as my vicarious second home, I have found many colleagues there who share my interest in the social, economic and political implications of the great human enterprise known as science. The kinds of ideas and considerations exposed in this book are at once focused on Australia, yet profoundly global in their significance. The issues raised here ring alarms just as relevant and important for Australia as for Canada and the rest of the world.

Discoveries and techniques for manipulating the molecules that specify the heredity and forms that all life takes represent truly revolutionary powers. The speed with which these capabilities have been acquired and the immense consequences of applying them demand widespread public examination and discussion. Having spent a quarter of a century as an active geneticist, and having been fortunate to have co-authored the most extensively-used genetics textbook in the world (Griffiths *et al.* 1996), I have nevertheless been concerned about the premature rush to exploit genetic technology—as well as the effusive announcements of its potential benefits. It may be instructive to indicate why I, a geneticist, have been led to raise questions and concerns about this exciting activity being carried out by my colleagues and students.

THE HUMAN SIDE OF A GENETICIST

I was born on the west coast of Canada in Vancouver in 1936. I was the third generation of the Suzuki family in Canada and remember an early childhood full of happiness and promise. In 1942, my family was declared a danger to Canada, stripped of all rights of citizenship and incarcerated in a remote valley in the Rocky Mountains for three years.

In 1945, we were presented with two choices: renounce our citizenship and leave the country or leave British Columbia for an exile east of the Rockies. The crime we, along with some 20 000 other Canadians, had committed was the possession of genes that had originated in Japan. The Japanese–Canadian experience during World War II is seared into my psyche, the source of immense insecurities and self-loathing. It instilled an excessive drive to gain approval from Canadians of my worth and it has made me hypersensitive to racism and injustice. Impoverished by the war and exiled in Ontario, my family seized on hard work and education as the means of improving our lot. As a recipient of a large scholarship, I was able to attend an outstanding liberal arts college in the US where, for an honours biology curriculum, I was required to take a course in genetics. Under the spell of my professor, I fell madly in love with this incredible subject as he posed fascinating questions and then revealed how scientists had uncovered the answers with precision and elegance. I was hooked and dreamed of going on to become a practising geneticist.

SPUTNIK AND THE EXPLOSION OF SCIENCE

The year before I graduated from college, the Soviet Union electrified the world with its dazzling achievement of sending Sputnik into space. In the agonising months that followed, American attempts to duplicate the feat failed spectacularly. In contrast, the Soviet juggernaut seemed invincible as they launched a series of firsts—dog, chimpanzee, man, team, woman—into orbit. It became clear that the Soviet Union was highly advanced in mathematics, engineering and science and, in response, the US poured billions into its space effort and into building up science.

As a Canadian, I nonetheless became a beneficiary when I entered a US graduate school. It was a time of seemingly limitless opportunity for budding scientists. We were taught, and I believed, that science is the most powerful way of knowing, enabling us to examine and explain the deepest secrets of the cosmos. In our minds, nothing lay beyond the enquiring eye of scientists and, with more research, life for all people would get better and better. I didn't question this article of faith and taught it to my students in my early years as a professor.

But two things happened after I graduated in 1961 as a fully-licensed scientist that made me question my assumptions. When I began to teach, students would ask me questions that I couldn't answer so I had to read more widely in order to find answers. In the process, I learned that there were parts of the history of genetics that none of my professors had taught me. And, as the years went by and the scientific community exploded in number and vigour, it became clear that despite the remarkable discoveries being made, life was not getting better for most people on earth while the planet itself was beginning to show unmistakable signs of distress. Let me discuss these points in more detail.

GENETICS' DARK HISTORY

As a teacher, I was enthralled by the rapid progress being made in genetics and loved to share the excitement by speculating on the benefits that could be gained from our insights. Students would often ask me questions which I found that I couldn't always answer—like what about the moral and ethical implications of cloning, what is eugenics, who controls the application of genetic knowledge, what responsibilities does a practising geneticist have, and so on. As I began to read, I discovered to my horror that early in this century, geneticists, intoxicated with the powerful insights that they were acquiring in this infant field, began to extrapolate from studies of physical traits in maize and fruitflies, to behaviour and intelligence in people. In addition to promulgating the mistaken belief that such human attributes as drunkenness, sloth, criminality and deceit follow the same hereditary principles as feather pattern in chickens or seed colour in peas, geneticists began to write and speak of traits that were 'desirable' or 'undesirable', 'superior' or 'inferior'. But these are value judgements, not scientifically meaningful categories.

In their exuberance over the burgeoning field, eminent geneticists, not third-rate scientists, made bold claims that applications of principles from their discipline could improve the human condition. The consequences of their boasts were not trivial. *Eugenics Acts* were passed in many countries that resulted in the sterilisation of people judged to be undesirable or inferior; American immigration restrictions were based on the notion that certain races have characteristics rendering them either preferable, or less desirable; many States enacted anti-miscegenation laws to prohibit inter-racial marriage on the grounds that it leads to 'disharmonious combinations' in their children.

For me, the most shocking realisation was that the Holocaust was not the result of the ideas of a madman, but had been based on the grand claims of geneticists. After all, Germany had an outstanding community of geneticists who were at the cutting edge of the field and many of them pressed for the improvement of Germany's gene pool by eliminating undesirables and encouraging the best. I learned to my horror that Josef Mengele, the infamous doctor at Auschwitz, was a human geneticist who held peer-reviewed grants for his studies of twins in the death camp. And in Canada, transcripts of Parliamentary debates before and after Pearl Harbor show that politicians from British Columbia seized on the claims of geneticists as the rationale for eliminating part of the 'Yellow Peril'. The evacuation, incarceration and expulsion of Japanese–Canadians had been encouraged by the grandiose claims of geneticists! So two of the great passions in my life—the fight against bigotry and the study of genetics—intersected in a grotesque way.

What was most disturbing was that none of my professors had ever mentioned this part of the history of genetics. I had been taught science as a steady progression of insights by intellectual giants. It had never occurred to me that these heroic figures were fallible human beings, as competitive, jealous, ambitious and biased as any other group of experts. But equally disturbing was the fact that in many fields, but especially genetics, scientists in the 1960s and 1970s were making the same kinds of claims that history should have warned them against. Molecular biologists, understandably thrilled to have acquired incredible manipulative capabilities with DNA, were claiming to have their hands on the levers of life so that hunger, poverty, deleterious mutations, disease and other afflictions were potentially controllable.

SOCIOLOGY OF GENETICS

The sociology of the scientific enterprise has changed radically since I graduated. In 1961, we were proud of doing basic research, seeking answers to questions that had no immediate application but added to our understanding of how life works. At conferences, scientists shared ideas openly and freely, co-operated on projects, respected areas already being investigated. There was excitement and a feeling of community. Today the incredible techniques for manipulating DNA have created a rush to patent DNA sequences and genetically-altered life forms and realise massive fortunes. Where once scientists sought answers to fundamental questions, they now seek venture capital investors for their biotechnology companies. Needing to produce ever more valuable output, researchers have become secretive, competitive and close-mouthed as the 'bottomline' has become driven by economics, not curiosity. In the 1920s and 1930s, scientists confused their personal values and beliefs about criminality, intelligence and abnormal behaviour with scientifically-verified fact. Today, scientists fall into a similar trap when they confuse *correlation* with *causation*. Thus, as geneticists report on the correlation of a trait or characteristic with a specific locus or DNA sequence, they readily slip into an assumption of a causal relationship. This was striking when researchers reported an 80 per cent correlation of alcoholics and non-alcoholics with different alleles of the alcohol dehydrogenase gene. As soon as the report was made public, scientists and reporters alike discussed the allelic differences as the cause of the dependence or non-dependence on alcohol. This is a serious mistake. It is like concluding that the high incidence of yellow stained fingers and teeth among lung cancer patients proves that stained teeth and fingers cause the disease.

THE HUMAN GENOME PROJECT AND ITS REAL IMPLICATIONS

The application of molecular techniques to the rapid sequencing of entire genomes is an astounding feat. I certainly never dreamed when I graduated that I would live to see the acquisition of such abilities in my lifetime. Biologists have designated the sequencing of the entire human genome with religious overtones. Nobel laureate Walter Gilbert, for example, declared that the project is the 'Holy Grail' of biology. The benefits of this project are widely claimed to be insights that will lead to understanding and cures for virtually every human condition for which there is a hereditary component.

The example of sickle cell anemia should serve as a warning for those making such bold claims. The genetic and molecular determinants of sickle cell anemia have been well known for over four decades but an effective treatment eludes us because of our vast ignorance about the human body, its components and their functions. While the completion of the Human Genome Project will be a monumental technological achievement, it will merely provide a linear sequence of instructions that must be read and transformed into a three dimensional entity that changes over time. We are a long way from knowing much about that.

As scientists exhaust the medical conditions for which sequence correlations can be screened, they will undoubtedly turn to other categories of human 'problems'—criminality, chronic poverty, alcoholism, homosexuality, intelligence levels, and so forth. And here, while the consequences of confusing correlation with causation are immense, the temptation to do so will also be strong. It seems far easier to blame the victim for his or her condition than to consider the impact of the social institutionalisation of racism, discrimination and inequities.

TECHNOLOGY—ITS BENEFITS AND COSTS

Biotechnology has vast social, economic and political ramifications and by focusing on this discipline, we gain insights into implications for technology as a whole. While scientists acquire knowledge about the world that is fragmented into bits and pieces, their discoveries can be applied with enormous impact. In chemistry, a host of new chemicals has been used for the production of synthetic materials, nerve gas and pesticides, for example. Discoveries in physics enabled the creation of X-ray machines, nuclear weapons and plants, radar, radio telescopes and compact discs. In medicine, researchers have provided antibiotics, nuclear imaging and oral contraception. The benefits have been immediate and obvious because we deliberately design technologies to perform specific functions. However, because our knowledge of the real world in which those applications will be used is so fragmented and incomplete, we have no capacity to predict beyond their immediate utility to their long-term consequences.

We have had enough experience with technology to know that however beneficial, every technology, from bows and arrows to antibiotics, nevertheless has costs. But our ignorance ensures that there will always be deleterious effects that cannot be predicted beforehand. Take for example, pesticides. When DDT was shown to be a powerful insecticide, the benefits were immediate and obvious—pests such as mosquitoes and weevils could be killed, and farmers and chemical companies would make money. At the time, geneticists already knew that resistant mutants would be rapidly selected so that more or different chemicals would soon have to be applied as the pests became resistant. And ecologists could have pointed out that insects are the most numerous group of animals on earth and are an essential part of all ecosystems. Only one or two of every thousand insect species are pests to humans and it is folly to spray broad spectrum insecticides simply to eliminate the small number of pests. That would be like killing everyone in New York city to manage crime in the city! But no one could have predicted that when a molecule like DDT is sprayed at concentrations of parts per million, the molecules would be consumed and concentrated up the food chain by several orders of magnitude. That is because biomagnification was unknown to scientists as a phenomenon and was only discovered when bird populations were reduced significantly.

Over and over again, our intoxication with new technologies has been dampened by the discovery of far reaching, unanticipated, effects. Biotechnology will be no different but because we are manipulating the very blueprint of living things, the consequences will be monumental. However, those who express concerns about the hazards of widespread use of new techniques or products are placed in the impossible position of having to define what the dangers might be. Because we know so little, it is not possible to know what deleterious effects there will be even though we can say with certainty that there will be hazards. Unfortunately, our technological activities are 'managed' on the assumption that, unless a concrete hazard can be specified, one does not exist. It is that very way of thinking that has thrust us into a global ecological crisis.

SCIENTISM

The remarkable success of science when applied by the military, industry and medicine explains the pre-eminence that the scientific community has acquired in this century. There is a widespread belief that science is the most profound way of knowing and allows us to know the 'true' state of the world. 'Scientism' is the belief in the authority of science and is widely held not only by scientists, politicians and businesspeople, but also by the general public. It is a dangerous myth and scientists themselves should be the first to know and demystify it. Consider this.

I graduated as a fully licensed geneticist in 1961, full of excitement at the revolution in genetic thinking that was then going on. I was enthralled and arrogant. I felt scientists were pushing back the curtains of ignorance and understanding the deepest mysteries of biology. Today when I tell students what we believed about the structure of chromosomes, how genes were organised and the mechanisms of development and differentiation, they laugh in disbelief. Considered in 1997, the leading ideas of 1961 seem naive and absurd. But students are then stunned when I remind them that twenty years from now when *they* are professors, the hottest ideas of 1998 will seem as ridiculous to their students. Why then, I ask them, are you in such a hurry to apply the latest ideas? The nature of the scientific enterprise is that most of our current ideas will ultimately prove to be wrong, irrelevant or unimportant. This is not a denigration of scientists—science advances by having current ideas disproved. The very nature of the scientific enterprise undermines scientism as a belief. It is sobering to realise that as scientists rush to patent and capitalise on their incremental insights, the rationale for their applications will undoubtedly be wrong, wide of the mark, or trivial.

Science has occupied most of my adult life and has provided me with the happiest times and most exhilarating moments. I am proud to have been an active member of the scientific community and to have contributed in a small way to the body of scientific knowledge. It is because I love science and care passionately that I am compelled to share my concerns for the lack of understanding of the history of science, the nature of the scientific enterprise, how science differs from other ways of knowing, its power and its limits—and the responsibility of its practitioners. I am honoured to be part of this book which dares to reflect on the implications and the untrammelled benefits of the genetic revolution.

Part 1

Setting the agenda

1

Bio-Utopia: the way forward?

RICHARD HINDMARSH, GEOFFREY LAWRENCE
AND JANET NORTON

Genetic engineering is changing the world. Emerging in the early 1970s—after a gestation period of forty years—it is the most powerful tool yet developed for the manipulation of Nature. Also known as recombinant-DNA (r-DNA) technology, genetic engineering alters radically an age-old technology known as 'biotechnology'. 'Bio' is derived from the Greek *bios*—meaning 'life'—and people have for many thousands of years used varying technologies to modify, for human advantage, life forms and processes. Fermentation of beers and wines are but two obvious examples. Over time, new techniques for manipulating life have been added via industrial microbiology, biochemistry, chemical engineering and tissue culture.

Today, through genetic engineering, *modern* biotechnology (henceforth used interchangeably with biotechnology and gene-biotechnology) attracts increasing government and corporate support and is set to make a major impact on the twenty-first century. Indeed, with many biotechnology programmes underway—including the genetic mapping of human, plant, animal and microbe genomes—some believe that the Age of Biology is upon us (see Venter and Cohen 1997). The economic impact alone will be immense. According to bioindustry analysts, by the year 2025 some 70 per cent of the industrial economy and 40 per cent of the entire global economy will have, at its base, some form of biotechnology (McAuliffe and McAuliffe 1981; RAFI 1989, p. 8).

The radical departure of r-DNA technology from previous biotechnological manipulations of Nature—such as artificial insemination, *in vitro* fertilisation, or plant breeding through traditional hybridisation—is illuminated upon by David Suzuki and Peter Knudston (1988, p. 116).

> In 1973, two decades after Watson and Crick published their revelations about the architecture of the DNA double helix, a historic event took place that marked the beginning of modern genetic engineering. Herbert

> Boyer, a researcher at the University of California, and Stanley Cohen, at Stanford University, succeeded in ferrying a recombinant DNA molecule containing DNA sequences from a toad and a bacterium into a living bacterial cell. There, to almost everyone's astonishment, the foreign toad DNA was copied and biologically expressed in protein. For the first time, scientists had choreographed genes from the cells of an evolutionarily advanced species to dance in the cells of a distantly related species.

The techniques of genetic engineering thus render DNA (deoxyribonucleic acid—the carrier of genetic information) subject to direct human access and control. Based on 'reweaving the threads of life' through cellular mechanics, the so-called 'blueprint of life' can be redesigned at the 'inside'. Bioscientists can 'unzip' DNA and inactivate or delete it from cells, or recombine the material in cross-species DNA transfers, producing novel organisms hitherto unknown in Nature. If the social power and influence of gene-altering proponents allows them to bioindustrialise the world, as they seem poised to do, dramatic change may follow:

> . . . genetically modified crops, livestock and pastures will be leading contributors to the economy . . . Gene and embryo technologies will tailor animals to their environment as well as to highly demanding markets . . . deliver designer enzymes to create high-value rural products . . . improve food quality, flavour and safety . . . yield superior breeds of leisure animals, from aquarium fish to racehorses . . . give rise to modified animals whose organs can be transplanted into humans . . . Medicine will be revolutionised . . . gene tests will determine the likelihood of an individual developing diseases and identify them, as well as behavioural and physical tendencies. Recreation will be enhanced by parks, golf courses and gardens containing wholly new organisms—plants with long life, unique colours and grasses that don't need watering or mowing . . . Huge off-shore cages will be built for the rearing of fish . . . fish that grow faster, resist disease and have better nutritional qualities . . . (see Cribb 1994, p. 9).

These are features of the bio-Utopian future envisioned by scientists from Australia's Commonwealth Scientific and Industrial Research Organisation (CSIRO). 'In laboratories across the nation', they claim, 'the foundations of tomorrow are already being laid' (see Cribb 1994, p. 9). With biological frontiers worldwide being recast by genetic engineering, these may be optimistic forecasts, but they are not science fiction.

Newspapers, magazines, science shows, and Internet sites now constantly depict heady, revolutionary shapshots of the 'breakthroughs' of the new biology—from scientific claims for 'big effect' genes that code for obesity, Alzheimer's disease, or for personality; to the engineering of plants to adapt to global warming. Nothing seems impossible. The rate of discovery—which is fast outstripping other areas of scientific endeavour—parallels the zeal of biotechnology's proponents to promote their science as the key to human progress. Civilisation, finally, will be able to control its biological destiny. By escaping from their genetic straightjackets and the constraints of Nature, people will find a new kind of freedom—a freedom to overcome disease and hunger, to have an improved standard of health and, of course, to live longer.

Yet, sometimes developments from the laboratory make us sharply pause and deliberate upon the bio-Upotian dream—and to ponder just what is going on. Is genetic engineering really rational science or are the boundaries between reality and science fiction becoming blurred? For some, the dominant image of the scientist in twentieth-century fiction and film—Frankenstein—comes to mind. Frankenstein signalled the ethical dilemmas and inherent dangers of reconstructing nature through a technology of living substance, as literary historian Roslynn Haynes describes:

> In scientific terms, the creation of the Monster is a brilliant achievement; yet Frankenstein's horror begins at the precise moment when the creature opens its eye, the moment when for the first time Frankenstein himself is no longer in control of his experiment. His creation is now autonomous and cannot be uncreated any more than the results of scientific research can be unlearned, or the contents of Pandora's box recaptured (Haynes 1994, p. 97).

For others, H.G. Well's novel *The Island of Dr Moreau* comes to mind with the extraordinary exploits of Dr Moreau's vivisection of animals to make them human-like. Yet, for others, the new biology resembles that of Aldous Huxley's *Brave New World*, which describes a society in which humans are 'decanted' not born; where individuals are grown in bottles with the required characteristics to suit the life that they will be expected to lead 'as Alphas or Eposilons, as future sewage workers or [as] future . . . Directors' (Huxley 1932, reprinted 1977, p. 22). How far away is such a world? Huxley himself reflected:

> All things considered it looks as though Utopia were far closer to us than anyone . . . could have imagined. Then, I projected it six hundred years into the future. Today it seems quite possible that the horror may be upon us within a single century (Huxley 1946, reprinted 1977, p. 14).

His words seem prophetic. In 1997, the genetic production of a headless frog embryo from a UK laboratory demonstrates just how close bio-Utopia is coming:

> London: British scientists have created a frog embryo without a head, a technique that may lead to the production of headless human clones to

> grow organs and tissue for transplant . . . such as hearts, kidneys and livers in an embryonic sack living in an artificial womb. Many scientists believe human cloning is inevitable following the birth of the sheep Dolly, the world's first cloned mammal . . . [P]eople needing transplants could have organs 'grown to order' from their own cloned cells (*Sydney Morning Herald* 20 October 1997, p. 11).

Yet, instead of addressing the issue of the morality of altering life that results in such bizarre and unnatural outcomes, the research was ethically justified as being in the services of humanity. After all, partial embryos—the result of the genetic manipulation of the egg to suppress development of the head—may not even technically qualify as embryos because they lack a brain or central nervous system. It was thus argued that the beauty of such developments was that because the donor is never sentient to begin with—'what could be the harm?' (see Morton 1997, p. 798). If the morality question were indeed reduced to a mere technical consideration, then growing partial embryos to cultivate customised organs could potentially bypass legal restrictions and ethical concerns. Yet, considering the outrage from certain quarters that the headless frog embryo instantly attracted—and before it concerns over 'Dolly' the cloned sheep and the possibility of human clones (see *Nature* 1997, p. 656)—it is hard to imagine that only technical considerations will suffice. To Oxford University animal ethicist Professor Andrew Linzey, 'it [was] morally regressive to create a mutant form of life' such as the headless frog embryo, and the work amounted to 'scientific fascism because we would be creating other beings whose very existence would be to serve the dominant group' (*Sydney Morning Herald* 20 October 1997, p. 11).

This ethical concern—as well as others about manipulating DNA—are increasing the intensity of debate about the purpose and outcomes of molecular genetics and genetic manipulation, as well as about associated biotechnologies such as tissue or cell culture.[1] Indeed, they highlight a dualism that has emerged in this debate which centres vividly upon the perceived benefits and costs of genetic engineering. This dualism, Wheale and McNally (1988, p. xvi) refer to as the 'Janus-face' of biotechnology:

> The genetic engineer, like a contemporary Daedalus, claims to be providing society with a vast range of innovations . . . On the other hand, as a result of the application of genetic engineering, the triggering of catastrophic ecological imbalances by the release of novel organisms into the environment, the creation of new agents of biological warfare and the increased power to manipulate and control people, may . . . become realities in the near future.

To the proponents, genetic engineering is a revolutionary scientific innovation that can provide new opportunities for improved health, protection from infection, control of diseases in crops, animals and humans, and for economic return. Some risk must be borne by society for technological progress. With a burgeoning world population, with increasing environmental pollution and social crisis, and with people's desire to live longer—the risk is worth taking. For the antagonists, though, science is indeed proceeding down an unknown path into an anti-Utopia or dystopia where *algenists*—latter-day alchemists—attempt to subordinate and reconstruct Nature (see Rifkin 1984). The result could easily be ecological destruction on a scale never before seen. Given the genetic combinations now being concocted by the genetic engineers, as well as our extremely limited knowledge about the workings of natural systems, there is little way of knowing what longer-term risks human beings and the wider ecosystem face.

Others are concerned about the social, political and economic structures into which the new technology is embedded. As much of the new biology is driven by the motive of corporate profit, how much of the research will genuinely produce widespread benefits? People fear that questions of ethics of the appropriateness of the technology for a sustainable future, and of whose ultimate interests are being served, will be swept aside in the race for commercial rewards. They argue that—as was shown by the preceding Green Revolution of industrial agriculture (see George 1976; Hobbelink 1991; Shiva 1997)—most of the benefits will go to those with power and wealth, not to those economically disadvantaged who are often touted as the eventual beneficiaries of the new Gene Revolution.

In response, the proponents argue that a commercial edge is necessary for product development and for the widespread distribution of benefits from the new products: people risking money should rightly claim a return on their investments—something which, after all, will enhance their capacity to innovate for the future. The marketplace will determine what is developed, what produces profits and what fails. Moreover, the technology is viewed as being as safe as possible. The counter arguments, however, about safety have been—and always will be—'we have this area of great uncertainty', and 'what about the broader needs of the wider community'. To sum up, the proponents believe society must take a calculated risk with biotechnology: the opponents believe that the risk is simply not worth taking.

Altered species being developed for the twenty-first century marketplace represent a cornucopia of novel medical, meat and dairy products, and fruit and vegetables in shapes, colours, flavours and textures never before seen. Already on the market, or in the research and development (R and D) pipeline are: the onco-mouse—a research animal with a cancer gene inserted into the animal's DNA; 'Dolly' the cloned sheep and 'Polly' the cloned lamb both emerging from laboratory experiments in 1997; 'Mr Jefferson' the cloned calf from US experiments in 1998; edible plant vaccines; bioengineered plants that tolerate salinity, acidified soils, and resist disease, ever-hardy insects and viruses; non-germinating wheat and rice seeds; genetically-engineered animal organs (known as xenotransplants) designed for human transplantation; non-ripening 'anti-rot' tomatoes; faster growing crops; biodegradable, plastic-producing, plants; purple carnations and blue roses; faster growing pigs and fish; and cows encouraged to produce more milk via injectable r-DNA bovine growth hormone, also known as bovine somatotropin (bST).

New transgenic breeds of sheep, goats and cows have been designed to secrete novel human proteins into their milk which can later be extracted and used in the pharmaceutical industry. The same animals can be programmed to produce nutraceuticals (higher-profit value-added food products) which—at least in theory—will result in healthier foods of greater nutrition. Another research project is to insert animals with human DNA to produce human blood plasma. These transgenic animals have been mechanistically renamed 'bioreactors'. They reside not on the farm but 'graze' at 'bio-pharming' companies like Genzyme Transgenics Corporation.

The proponents' promise for these bio-products—feeding the world's hungry, improving the health of all, conquering disease, creating new 'clean and green' industries for ecological sustainability—is naturally alluring. Yet, a darker scenario also exists. It is that of a bioindustrialised world dominated by a technical life-sciences elite (or bio-elite) whose corporate linkages ensure profitmaking via patents on life and biodiversity—the new 'gene-gold', and whose research concentrates on the symptoms of problems, not their causes. New food and health-care production systems based on a quick-fix approach will address the superficial 'needs' of a consumer society but fail generally to alleviate the growing social and ecological crisis which society currently faces. Indeed, the ecological crisis may worsen through the widescale release of transgenic organisms into the environment creating a new form of pollution—genetic pollution.

This bio-Utopia may also greatly disadvantage vulnerable segments of society through increased social control; through genetic discrimination (referred to in the 1997 genetics science fiction movie thriller—*Gattaca*—as 'genoism'); and through 'bio-colonialism'. It may become a world where there is compulsory genetic testing and DNA fingerprinting; where insurance companies or potential employers access human genetic information through DNA databanks; where women's control over procreation is further decreased by gene therapy and by the further promulgation of the myth of human perfectibility. It may be a world where horrifying new weapons of biological warfare emerge; a world where indigenous peoples' genes, and knowledge and ownership of their flora's genetic information, is appropriated by Western firms for genetic databanks; and where potentially dangerous r-DNA vaccines are tested in unregulated Third World countries.

Farming and food manufacturing systems may also be altered in a manner which disadvantages—through price, accessibility, or lack of choice—the world's poor and hungry. Altered genes—in the form of novel foods—may also alienate a growing environmentally aware consumer movement, one which may not tolerate genetic manipulation of either Nature or of food. If so, the agri-food industries—something in which people have tended, in past decades, to rely upon for the provision of healthy, nutritious foods—may become the site of heightened struggle, not only over the 'meaning' of food, but also over its composition and delivery. This will especially be the case if unlabelled genetically-engineered foods continue to be sold to unsuspecting customers who believe consumer sovereignty has been violated. Despite such unresolved questions about its promise, its ethical and scientific basis, political alliances, and social and environmental impact, global biodevelopment continues unabated.

GLOBAL BIOINDUSTRIAL DEVELOPMENT

Annual global investment in biotechnology is now in the vicinity of US$20 billion, of which around two-thirds comes from private business and one-third from governments—predominantly businesses and governments in the First World. While many companies, government agencies, and scientists aim to put molecular techniques to useful service, commercial imperatives shape and propel the technology. Handsome profits are envisaged. Within the European Union alone—which comes second only to North America in its number of biotechnology companies—an industry study has estimated the potential market for biotechnology-related products at US$278 billion by the year 2005 (*Agbiotech Bulletin* 1997).

Following the Boyer-Cohen r-DNA breakthrough of 1973, and the worldwide diffusion of biotechnology's commercial promise in the late 1970s, a wave of investment emerged to fund new biotechnology firms (NBFs)—which were usually commercial 'spin-offs' from universities. In the early 1980s corporate capital began to flood bioresearch institutes. Since then, NBF take-overs, corporate mergers, as well as the vastly superior economic power of corporate entities to invest in R and D and engage in patent battles, have acted in concert to narrow the competitive field (see Krimsky 1991; RAFI 1996a; Hindmarsh 1998). With some 2800 biotechnology companies worldwide, multinational corporations

(MNCs) now clearly dominate biodevelopment. MNCs presently control some 20–30 per cent of the fledgling bioindustry, a figure predicted to rise to 50 per cent in the near future.

In the top tier of life sciences conglomerates are Monsanto, Pioneer Hi-Bred, Dupont, Dow Chemical, AgrEvo, and Zeneca (GRAIN 1996), and at the forefront is Novartis—formed, in 1996, by the US$27 billion merger of Swiss giants Sandoz and Ciba-Giegy. Novartis is now the world's largest agrochemical company, the second largest seed firm, the third largest pharmaceutical firm and the fourth largest veterinary medicine company (RAFI 1996).

From the first field trial conducted in 1986 to 1995, more than 4000 field tests of transgenic crops had been conducted at more than 15 000 individual sites in 34 countries. Some 35 per cent (the majority) were for herbicide tolerance, including crops being developed to tolerate or resist a particular company's own herbicide. Crops receiving the most attention include soy beans, maize (corn), potatoes, melon/squash, tomatoes and canola, along with tobacco and cotton (James and Krattiger 1996). Australian biodevelopment is intricately entwined with global bioindustrialisation.

AUSTRALIAN BIOINDUSTRIAL DEVELOPMENT

Australia's early involvement with international r-DNA activities most markedly began when delegates from the Australian Academy of Science attended the 1975 Asilomar conference in California. That meeting provided a platform for molecular scientists from all over the world to defend their experimentation, and to reject any restrictive external interference in the development of bioscience (see Krimsky 1991; chapter 3 this volume). Asilomar provided a landmark opportunity to assert scientifically (and politically) the desirability of a 'biotechnology future' internationally, and for Australia.

Based on the not inconsiderable intellectual resources within the CSIRO and the university sector, bioscientific knowledge and the bioindustry have developed quickly in Australia. International connections have been enthusiastically sought as a means of both increasing the commercial equity in experimentation, and of gaining new and potentially lucrative world markets. By 1990, over one-third of the some 100 Australian firms involved in gene-biotechnology had formed strategic alliances with overseas firms, particularly with European, US and Japanese firms (Scott-Kemmis *et al.* 1990). Another 30 firms were also either largely funded by, or were subsidiaries of (or were biotechnology divisions within), large corporations. More specifically, some 36 Australian firms are now considered to be fully-fledged biotechnology companies because of their direct focus on genetic engineering, monoclonal antibody technology, engineering of biologically active materials (such as proteins), vaccines, *in vitro* cell manipulation, micropropagation, and sophisticated biochemical, microbiological and immunological technologies (ABA 1993). Currently, the federal government invests over A$100 million a year into r-DNA technology, with the CSIRO capturing about one-third of those funds. Private investment also exceeds A$100 million a year (see Hindmarsh 1994).

Overall, Australia's involvement in global bioindustrialisation clearly indicates that the nation is increasingly dependent upon corporate decisions which are taken abroad, and based on the interests of distant corporate mangers and shareholders. It is a position which has an element of neo-colonial dependency. Australian involvement centres upon its roles as a specialist service provider (of scientific know-how) to corporate interests; as a 'captive' technological client (both as adopter and developer) of the new biotechnological processes and products; as an active (if relatively minor) player and proponent in the global bio-policy network; and, as a platform and conduit to the Asia–Pacific region for both local and global biodevelopment (Hindmarsh 1998). Table 1.1 provides a profile of Australian bio-business and public sector interlinkages. It shows that many top life-sciences multinational corporations have extended their operations into Australia. Like the situation overseas, a Byzantine web of formal contractual obligations and informal connections has emerged between public sector research agencies (such as the CSIRO), universities, NBFs and MNCs.

PUBLIC CONCERN AND SOCIAL RESISTANCE

The failure to adequately address and resolve the broader environmental, ethical and social implications of genetic engineering activities has engendered a global campaign of criticism and public protest. Environmentalists, consumer groups, animal-rights activists, organic-agriculture advocates, food-trade organisations, and concerned scientists, feminists, religious groups, and Third World and indigenous peoples, have all entered the debate. Almost without exception these groups have indicated how they (or their constituents) will be disadvantaged by the application of r-DNA technology.

Public concern has emerged in a diversity of forms. Direct action has seen underground eco-activist groups like the 'Fiery Viruses' and the 'Snarling Spuds' launch attacks on laboratories, hothouses and test sites of r-DNA plants in the Netherlands and Germany (see Abbott 1996). So-called transgenic-crop 'trashing' spread to the UK and Ireland during 1997, and then onto Australia where a consumer group 'Mothers

Table 1.1 Multinational corporate involvement in gene-biotechnology in Australia

MNC	Parent	Business activities	Australian operations
Monsanto	USA	Agrochemicals/ seeds	Agreements with CSIRO, and field-trials (glyphosate-resistant canola).
Lubrizol	USA	Chemicals/seeds	Local research through subsidiary Agrigenetics.
Pfizer	USA	Seeds	Received approval in 1990 to market GE microbes for cheese production.
Upjohn	USA	Human/animal health care Agronomic/veget-able seeds and speciality chemicals	Takeover of Perth-based Delta West Ltd for $A23M (1991).
Coca Cola Amital	USA	Food processor	Venture with CSIRO—virus-resistant potatoes
Johnson & Johnson	USA	Pharmaceutical	$25M joint venture with the CSIRO. Investment in Pac Bio. Owns Tasmania Alkaloids P/L, and Janssen-Cilag P/L.
ICI	UK	Agro/chemicals/ seeds	Contracts and joint ventures with universities and CSIRO. Acquired Aust. fertiliser company Incitec, seed company Pacific Seeds, and Monoclonal Aust. Ltd.
Unilever	UK/Dutch	Food processor/ seeds	Unifoods field trial for gene-altered tomatoes. Owns Oxoid Aust. P/L (pharmaceuticals/agbiotech).
Gist-Brocades	Dutch	Food and pharmaceuticals	Acquired Mauri Labs from Burns Philp.
Bayer	German	Pharmaceuticals/ agrochemicals	Owns Miles Australia P/L.
AgrEvo (Hoechst and Schering)	German	Agro/chemicals	Owns Regulin, owns Biotech Aust—vaccines and biopesticides. Field trials of herbicide-resistant canola.
Hoffman-La Roche	Swiss	Pharmaceuticals/ agribusiness	Owns Western Biotech Ltd (algae-biotechnology).
Sandoz (Novartis)	Swiss	Pharmaceutical/ seeds	Joint venture with AMRAD.
Suntory	Japan	Flowers	Joint venture with and 15% equity in Florigene.
Rhône-Poulenc	France	Agrochemical/ seeds	Took over Auspharm Internat. Ltd for $18.7M. 5 year collaboration with AMRAD (from May 1997).
Limagrain	France	Seeds	$25M joint venture with CSIRO to produce transgenic virus-resistant seeds. Took over research divison of Aust. company Ag-Seed.

Sources: Australasian Biotechnology, Scitech, various.

Against Genetic Engineering' claimed to have uprooted an experimental plot of genetically-engineered canola plants belonging to AgrEvo located at a field-trial site at Gatton outside Brisbane (AgrEvo subsequently denied any damage to the crop trials—see Roush 1997b).

In contrast, at the policy level, leading groups—such as the US Union of Concerned Scientists, the Greens in the European Parliament, and the Australian GeneEthics Network—lobby for strict regulation and attempt to raise public awareness and open debate. Internationally, many Green groups advocate a moratorium on environmental release of transgenic organisms, oppose herbicide-tolerant plants, and—together with a vast array of public interest groups worldwide—condemn the patenting of life-forms. A catalyst boosting the oppositional movement came in late 1996, when anti-genetic engineering group, the US Foundation of Economic Trends, joined with Greenpeace to initiate a global campaign to boycott Monsanto's 'Roundup Ready' soybeans (see *Nature Biotechnology* 1996; chapter 14 this volume). Almost instantaneously, the Pure Food campaign attracted support from more than 500 organisations in over 75 countries.

In response, a powerful proponent counter-campaign to align global society with a bio-Utopian future has been forged. Authoritative reports like *Agenda 21*—from the United Nations Conference on Environment and Development—aim to popularise biotechnology as a core solution to the world's environmental and social problems without, paradoxically, addressing most issues raised by concerned Third World, and environmental, groups. Another tack is to depict critics as scare-mongers (see Playne 1998, p. 39), extremists, zealots, neo-Luddites, as being unrepresentative of the wider public, or as working against economic progress (see Plein 1991). A 'softer' line is to suggest, in somewhat condescending fashion, that while criticism might be well-intentioned it is nevertheless ill-informed and wrong. In these ways, critics are construed as working against the interests of modern society. This obviously attempts to marginalise their concerns. The public is further told that the critics simply fail to understand the beneficial nature of genetic engineering, and that if they could, they would see it as a proven, safe and reliable technology. Consumers and critics are positioned, in short, to be scientifically naive.

Most often, when criticism is levelled at new mega-technology, the result invariably has been a vigorous and uncompromising defence of the technology. This seemingly 'siege-mentality' of technical elites in the face of opposition is designed to place a fetter on discussion and debate. Rather than confronting public concerns about technology in a constructive manner, the approach of scientists and industry is most often to manouvre to marginalise those concerns by either dismissing them or by *reassuring* the public that nothing is untoward. Such a response hough is embedded in an even deeper malaise—one marked by the social insulation of science and a technical reductionism which drives much industrial innovation (Wynne 1983). Social and environmental problems then arise when socially-insulated innovation is taken out of its vacuum and plunged into the bustle of the real world.

Such problems are highly apparent with genetic engineering innovation. As a result, and in response, political uncertainty and public distrust has emerged in the form of conflict and social resistance. With many people bewildered by some of the emerging gene-altered life-forms, questions are being asked about just *what* is going on. Proponents are facing increasing difficulty in their attempts to dismiss or absorb mounting and accumulating public concerns. Enter the new weapon of the so-called 'public education' programmes. These have been developed to persuade a sceptical public of the virtues of genetic engineering (see Hindmarsh 1996; chapter 3 this volume). A plethora of pro-biotechnology information packages is now visible on the Internet and in mailouts particularly targeted to younger, potentially more accepting, high-school audiences.

PUBLIC DEBATE

In an open, democratic society many people would challenge the limiting of public involvement in decision-making processes concerning the technological direction of society. Yet, it is the view of many scientists and technical experts that the public is inexpert and its views sometimes alarmist. Some consider that public criticism and concern is thus unwarranted, that it restricts scientific freedom, threatens scientific 'neutrality', damages scientific and economic progress, and is a harbinger of the political control of science (see Holman and Dutton 1978). What is forgotten here, though, is that technology is an inherently social construct—embedded in structures of social power which, in turn, reproduce systems of domination, shape agendas for change, direct society's preferences, and determine economic outcomes. In this context, science is not neutral or apolitical, and scientists are not 'free' and 'independent' but are part of a system in which class, gender, racial, ethnic and other social relations affect, and are affected by, scientific 'progress'.

Some have responded to such a realisation by proposing that matters scientific be removed from public debate. For example, genomic scientists Venter and Cohen (1997) propose the establishment of a worldwide upper chamber of parliament to address the ethical issues of genetics research. The parliament would be 'a deliberative body of experienced scientists and philosophers . . . to advise decision-makers in business and politics' and to 'inform the public of what is at stake in a given scientific advance and propose solutions' (Venter and Cohen 1997). While this proposal is well-meaning it overlooks the point that scientists and philosophers bring with them particular, minority views of the world. Where is the crucial role of the public's value systems in this top-down approach to directing technological change?

In democratic society an overwhelming case exists for public participation in matters which will impact upon people's lives and on the biosphere. First, the public has a right to know exactly what is going on and to participate in major social and environmental developments—including the development and application of technology. Second, public participation is certainly called for when it is understood that, through its taxes, the public provides money for (bio)technological experimentation and must also bear the consequences of its development. Third, effective representation of the public interest better assures the safety of commercial products and helps to anticipate—and therefore potentially to avoid—failures of science and technology (see Holman and Dutton 1978; Rissler and Mellon 1990). The latter aspect is, of course, essential for a transition to ecologically sustainable development (ESD)—a national strategy that Australia has adopted to attain long-term social and economic development which sustains the natural environment and promotes social equity. Finally, public participation is warranted because society's traditionally-perceived option of depending on the 'objective' analysis and expert opinion of scientists is undermined by the commercial and social power imperatives of (bio)technology. As is the case with most of the significant issues faced by society, public participation is essential if we are to avoid the potential abuses of power by those with vested interests (see Mitcham 1997).

These arguments, and the ESD imperative, require that other values—those of the public at large, those of groups and individuals affected by policy, and those of other species and of biodiversity—must be taken into account alongside considerations of economic benefit. It is no longer appropriate—if indeed it ever was—to leave the decisions to a technical elite, or to an amoral global marketplace (see Moser 1995). Participatory technology can help ensure that responsible scientific and technological decision-making occurs at all stages of technology choice and implementation.

At present, however, the Australian public, and the international public, has restricted knowledge of the wider social and environmental impacts and ethical implications of the 'bio-revolution'. Yet, this is a time in which Australians—just as with citizens overseas—are becoming increasingly worried about genetic engineering and its potential effects (for example, see discussion in chapter 13).

Though some sentiments have been expressed (see, for example, Butler 1997, p. 775), governments throughout the world appear to be doing little to address the public's concerns or realistically to address the need for public debate, even as key communicators, public interest groups, educators and consumers are increasingly expressing their desire for informed, accessible, information. People are demanding comment on the broader issues—not only to enhance public awareness and participate in the debate, but *just to know what is really going on.* Currently, little public information—except for the glowing accounts of proponents—exists about the Janus-face of biotechnological change. At the very time that knowledge needs to be promulgated, there exists a public information vacuum.

THE WAY FORWARD: THE PURPOSE OF *ALTERED GENES*

Altered Genes is the first Australasian book that presents a critical perspective of the many important issues raised by modern biotechnology, and which places those issues in the local context. With the highly controversial and threshold gene-biotechnologies and products now beginning to arrive in the marketplace, there is an urgent need to stimulate open and informed community participation. If we are to move toward an ecologically sustainable future, we must decide on where biotechnology 'fits'. It is in everybody's interests to discuss and to decide upon the desirability of the genetic engineering enterprise and its bio-Utopian agenda. While the 'promise' of biotechnology has certainly been presented to the public, the broader issues and potential 'costs' remain under-presented. With social resistance to gene-technology growing, in the face of PR campaigns promoting gene-technology, there is need to strike a balance. The public needs to be drawn in from the 'outside' to become aware of the whole debate. As many of the contributors to this volume indicate, the public desires to increase its understanding and awareness of the issues that the new technology poses. To do so, the public must recognise that bio-experimentation and commercialisation is embedded within the wider socio-economic, political, moral, cultural and ecological frameworks of contemporary society.

This book presents a range of issues that need to be brought to the public's attention. The viewpoints of its contributors, however, are by no means uniform in outlook. Some contributors see a 'brilliance'—though a cautious one—in the promise of gene-technology and believe that the genetically-engineered baby should not necessarily be thrown out with the biotechnological bathwater. Some see merits in some of its applications, but not in other ones. Other contributors find genetic engineering abhorrent, or inherently flawed, and think that the grand gene-altering experiment should cease. Some believe it would be better to consider other, alternative, approaches that work *with* nature and community—natural therapies, organic agriculture, and the like—approaches gaining in popularity both overseas and here in Australia. Yet other contributors display no strong position but raise important issues which society must face.

All contributors, though, would appear to agree with two propositions: first, there needs to be thorough public debate; and second, it is our right and duty as public academics and citizens to raise criticisms about—as well as to challenge the assumptions of—those preparing the world for a very different bio-future. The views of the contributors challenge the simplistic impressions that the lines of debate are clear and that the best way forward is to join one or other 'camp'—to adopt an 'us or them' approach. We recognise and applaud the diversity of opinion among the bio-critics, and we understand that there is—in parallel—a diversity of views among the bioscientists and industry representatives. Through open community debate the critics and proponents, and the public, stand to gain a greater understanding, awareness, and appreciation of the issues and problems surrounding the move towards bio-Utopia.

THE CONTRIBUTORS' VIEWPOINTS

The book has been divided into three sections, reflecting the major ideas and concerns of the contributors. Part 1 provides a framework for understanding the pervasiveness of biotechnology in public discourse and in practice, from the social to the ecological.

For Tiffany White the modern-day Lois Lane can never enjoy the journalistic licence of her fictitious counterpart from the Superman comics. The Lois of the real world would be hamstrung by deadlines, reliant upon the news and PR releases of companies, politicians and scientists, and would not be encouraged to venture too far away from the newspaper's orthodox position on science and technology. The result, as White shows in chapter 2, is the promulgation of pro-biotechnology information where the media act as legitimisers of current policy, failing to incorporate alternative perspectives. The bioscientists and the biotechnology industry have a secure, authoritative, position in society. Their news releases are often designed so their content can be uncritically reproduced by journalists who are unwilling or unable to spend the time to 'test' their assertions or assumptions. In White's view, science journalists in the Australian commercial media need more support from their editors and news organisations if they are to succeed in the increasingly important task of engendering public debate about topics such as biotechnology.

In chapter 3, Richard Hindmarsh draws upon the sociology of social power relations to analyse a tactical bio-policy network which emerged in Australia from the early years of the 1970s. This network continues today (with interlinkages of the CSIRO, universities, industry, government departments and various advisory bodies), issuing strong support for genetic engineering experimentation and biodevelopment. Drawing upon the work of Bruno Latour, Hindmarsh follows the members of the bio-policy network across the terrain of struggle over genetic engineering at two 'fronts of action'—regulation control and information control. Not seen in any simplistic conspiratorial manner, *bioscience in action* reflects an elite 'business-as-usual' and top-down approach to 'managing' democracy in order to ensure unhampered bioscientific progress, especially that linked to big business investment. Yet, despite the attempt to contain public debate and full knowledge, public concerns and the mobilisation of environmental and consumer groups have grown. In 'counter-attack' to this, the bio-policy network acts to manufacture public consent through 'education' programmes that seek to 'enrol' the public's acquiescence to genetic engineering. In that process, important community concerns about genetic engineering and increasing calls for public scrutiny and control of regulation, and for open public debate and information, are submerged in the ensuing environment of weak regulations and lack of information.

In chapter 4, Jean Christie examines the geography and politics of biodiversity and biotechnology worldwide, and places Australia in this context. She points out that biodiversity, the raw resource for all biotechnologies, is mostly to be found in the gene-rich, but cash-poor countries of Africa, Asia, Latin America—and in Australia. This, she argues, places Australia in a unique position: open for industrial exploitation like the developing countries of the 'South', yet politically in the camp of the industrialised 'North'. She contends that a desire by transnational industry to control genetic resources worldwide lies behind new industry trends, including bioprospecting, and the escalating use of intellectual property rights to secure monopolies over genetic resources, and that two international agreements—the World Trade Organisation's intellectual property agreement, and the Biodiversity Convention—facilitate these trends. Using four case studies, she looks at the impact of these trends especially on indigenous peoples of Australia as well as those of the poor countries of the 'South'. She raises these concerns for heightened public awareness and debate, and for a better outcome for indigenous peoples and biodiversity in the policy process.

In chapter 5, Peter Wills questions the assumptions upon which the assessment, safety and development of genetic engineering is based. He argues that modelling biological interactions as a linear process of genetic information—enshrined in the central dogma of molecular biology 'DNA makes RNA makes protein'—is a flawed representation of the relationship between genes and organisms. What is being missed is the fundamental circularity of genetic information processing, as well as the complexity of connections between biological entities and the environment. Overlooking these connections, molecular biologists and genetic engineers are showing themselves incapable of recognising the potentially dire impacts of the commercialisation and rapid and large-scale dissemination of genetically-engineered products into the environment. If this is allowed, then we can expect to perturb the dynamics of genetic change to such an extent that the changing ecological patterns which have emerged, survived, declined and re-emerged up until our current evolutionary epoch will be completely disrupted. Consequently, what is now urgently needed, Wills advocates, is an *indefinite* moratorium on environmental release of transgenic organisms.

Part 2 continues to explore issues of ethics of gene biology and does so by focusing on examples relating to human gene mapping, culture, and risk. In chapter 6 Robyn Rowland brings a strong feminist perspective to the topic of genetic engineering. Identifying the striving for 'perfection' as part of a narrow, masculine science, Rowland harks back to the eugenicists and their (misguided) attempts to create perfect human societies. Applied to human genome research, genetic engineering is viewed as a radical extension of male-controlled intervention in the lives of women. Women will be expected to conform to the notion of having 'perfect', 'unproblematic' children, and to carry genetically-manipulated embryos through to term, to deliver the resulting children, and to rear them. Many interrelated issues are raised in this 'dream of quality-control' including the control of the technology in corporate and state hands, the patenting of DNA, genetic discrimination, and social control. To resist, challenge or impede bio-Utopia from commodifying life, and using women's bodies as 'living laboratories', the empowerment of women and the community in regulatory and decision-making processes is vital.

In chapter 7 Nicholas Tonti-Filippini outlines problems with the bioethical prescription for the future that the United Nations 'Declaration on the Human Genome and Human Rights' represents. He also looks at the way the Australian biomedical establishment has responded to the ethical issues raised by the 'Declaration'. The problems posed include: Just what is the human genome project about? How is the individual protected? How is confidentiality maintained in relation to an individual's genetic characteristics? What if people face discrimination in relation to their genetic characteristics? And so on. The protection of human rights against the backdrop of an emerging era of bioengineering has not been properly addressed in Australia. Existing guidelines on ethical issues were drafted largely by scientists. Consequently, the guidelines fail to acknowledge or identify the many problems which are arising with the expansion of biotechnology, and which cannot readily be solved within a narrow scientific framework. Finishing on a positive note, Tonti-Filippini makes some important recommendations for change.

In chapter 8 the issue of genetic testing in the medical context is raised by Astrid Gesche. She summarises the problems associated with privacy and genetic information. Modern molecular genetics, she writes, allows us to test for both the presence of certain genetic aberrations and the likelihood of developing specific diseases in later life. But who, she asks, should have access to our personal genetic data, and what are the consequences of passing on this information? Should, for example, insurance companies or employers be afforded access to it? Gesche then proceeds to outline what privacy and confidentiality measures for molecular genetic information are currently offered by public and private Australian institutions. She cautions that, at present, data and information derived from medical genetic testing cannot be safeguarded adequately. She concludes by suggesting that any data transfer and exchange should be protected by legislation. Without it, the potential for misuse is simply too great.

Rosaleen Love takes flight from one prevailing media message that 'genes are destiny'. In chapter 9, she explores representations of the gene in the contemporary public culture of Australia, highlighting the so-called 'gene for death' and jokes about genes. By focusing upon the 'gene for death' she captures the paradox of the gene: that while life arises from genes, those genes also contain the potential for both deformity and death. Jokes about genes—a form of social resistance to genetic essentialism—provide a bridge between scientific and popular cultures, where non-scientists and scientists join in mocking current 'hyping' of the gene. Jokes point to the chinks in the 'world machine' view of life, the gap between the grandiose, hubristic claim that science has the answers and experience of the everyday world of increasing risk and hazard. A wider view of what it is to be human is provided by the immunologist Miroslav Holub in his reflections on the human genome and the human spirit.

In chapter 10 Stephen Crook offers a sociological analysis of biotechnology as 'cultural risk'. Crook argues that public anxieties about the risks of biotechnology are consistent with the pattern of an emergent 'risk society' as discussed by German sociologist Ulrich Beck. Drawing on the anthropology of Mary Douglas he emphasises the cultural dimensions of biotechnology risks. Biotechnology crosses fundamental boundaries between 'nature' and 'culture', threatening to pollute the sociocultural order just like witchcraft, or the eating of non-food animals. Risk-management is a major concern in risk societies and Crook identifies 'regimes' through which biotechnology risks can be managed. Because biotechnology is culturally so risky, the most significant regime for its management is ritualistic and rhetorical. The proponents of biotechnology mobilise 'reassurance strategies' centred on rhetorical moves designed to show that biotechnology is 'natural', 'traditional' and safe rather than 'artificial', 'new' and polluting. Crook illustrates this argument with examples drawn primarily from debates about genetically modified food. Although biotechnology's reassurance strategies are powerful and resonate with dominant assessments of science, their victory is not guaranteed. Anxious publics may distrust reassurances that they suspect are contrived solely to reassure.

Part 3 takes us to the farm, to the food table, and into the offices of those who are organising protests against Australia's biotechnology future. In chapter 11, Geoffrey Lawrence and Frank Vanclay examine biotechnology's rise as being based upon the public's desire to overcome environmental degradation, the farmer's needs to become more efficient and productive, and the food industry's requirement to become more cost-effective and competitive. The 'biotechnological solution' is one which science has projected for all three. This chapter examines the claims of those supporting the current trajectory of agro-biotechnology research in Australia and assesses the extent to which Australia's integration into the global food economy will be related to its utilisation of genetically-engineered products. It is argued that while the state views biotechnology as a 'saviour' for struggling farmers—and more generally for an agri-food sector facing enormous economic problems—any promise is falling far short of the reality. There are concerns that farmers may not make predicted gains from their genetically-manipulated inputs, that what constitutes a 'clean and green' agriculture may not equate with the novel products being developed by agribusiness, that release of genetically-engineered organisms may lead to more environmental problems, and that a gene-manipulated food industry may face consumer resistance—or outright rejection of many of the new products of biotechnology. There is simply no certainty that what bioscience is producing will be accepted by people—in Australia as well as abroad.

Anna Salleh provides a 'case study' of one of the genetically-engineered inputs to agriculture briefly described in the preceding chapter—transgenic cotton. She explains, in chapter 12, how the soil microbe *Bacillus thuringiensis* (Bt)—used sparingly for many years by organic farmers as an environmentally responsible pesticide—has been appropriated by the biotech industry in an attempt to overcome pest destruction of cotton crops. The gene coding for the microbe's insecticidal toxin has been transferred into the cotton plant. In theory, this means the cotton plant will be able to protect itself from pests; if pests try to eat the plant, they die. Moreover, it is said, farmers will be able to reduce the spraying of pesticides, making cotton growing a 'cleaner' system of production. Salleh then poses some probing observations about the product. The more Bt cotton is produced the more likely it is that insects will develop resistance to it. What if the new biological insecticide is eventually rendered useless by the transgenic experiment? What will the cotton producers turn to then? What will happen to the organic farmers, no longer able to rely on Bt for safe emergency pest control? For Salleh, the potential over-exploitation of Bt threatens to see the gene 'worn out' in a very short time—rendering suspect the promise that biotechnology will bring 'cleaner and greener' agriculture.

In chapter 13 Janet Norton describes the various genetically-engineered foods coming onto the market—proteins from gene-altered soy plants, cheeses made from genetically-modified starter cultures, flavours and sweeteners which have been bioengineered to exhibit certain properties, as well as the products from animals which have been treated with modified hormones or vaccines or which are, themselves, transgenic. While scientists argue that these products conform to previous farming and food processing practices, consumers appear unconvinced. Sociological research conducted both in Australia and abroad is reviewed in an attempt to establish what it is that consumers find problematic about the new foods. Food safety and food quality are but two of the issues identified in the surveys. Significantly, Norton's own research indicates that consumer resistance is not directed at the new food products intrinsically, but rather at the *type* of technology used to produce them. If conventional means are used, the products are acceptable to consumers. Acceptance drops, however, if they are to be produced using genetic engineering. It is concluded that while some consumer support is apparent where animal health may be enhanced (for example, through genes that code for resistance to blowflies) there is a general reluctance to ingest foods which have been genetically altered. Norton argues that any regulatory moves to ensure the labelling of all genetically-engineered foods would—from the viewpoint of consumers—be a highly desirable outcome.

In the final chapter, Bob Phelps picks up the campaign trail in Australia. As director of the Australian GeneEthics Network—a key lobby group with the goal of attaining genuine public control over all aspects of genetic engineering and its applications—Phelps is well placed to relate the campaign struggle in the policy terrain over genetic engineering. To introduce us to the strategies of genetic engineering proponents, and the associated social and environmental risks, issues and problems of genetic engineering, he selects the case study of Monsanto's soy 'gene-bean'—bioengineered to tolerate the company's own glyphosate herbicide 'Roundup'. With the use of soybean in a majority of processed foods—and with no apparent labelling in sight for products that might contain the 'Roundup Ready' soybean—Phelps' account offers a disturbing insight into the 'real' world of political manoeuvring being carried out to deploy gene-technologies without public knowledge or consent. Offering the alternative of sustainable agricultural systems to that of genetic engineering, Phelps highlights citizens' campaigns against genetic engineering, globally and locally, and concludes with the premise that, despite unbridled corporate power, we still have the opportunity to create a good legacy for future generations—a legacy that is free of a genetically-engineered dystopia.

NOTES

1 The 'case' of the headless frog embryo and the possible production of human body-part clones also raises the ethics of the lesser-publicised biotechnological technique of tissue or cell culture. A cell culture started from a single, multiplying cell is called a clone. It contains genetically identical copies of the single starting cell, although random mutation prevents all cells from being absolutely identical. The copying process is called cloning, and is applied whenever scientists want to obtain a large number of cells or organisms with the same features. Without cloning, genetic engineering would be useless for large-scale production processes. As implied in the report about the headless frog embryo, cell culture is now being extended from, for example, the growth of skin for skin grafts for burn victims, to the hoped-for production of organs for transplant. Cells, taken from a small number of cells taken directly from those destined to be recipients, are grown to become full sized organs in appropriately-shaped laboratory moulds. At present, in this field, only animal organs have been produced, but doctors anticipate that their dream of 'a shelf full of body parts' will soon be fulfilled (*Morning Bulletin*, 5 November 1997, p. 21).

2

'Get out of my lab, Lois!': In search of the media gene

TIFFANY WHITE

Lois Lane sits at her desk surrounded by paper, chewing off her lipstick and peering at the clock on the wall. She's doing the science round these days, and it's taken all morning to trawl the science journals for new discoveries and research to write up under her byline. There's a press conference at Vegtech this afternoon; they're releasing a genetically-engineered tomato. She rifles through her in-tray and finds a press release about it—it'll be firmer, redder, and last longer on the shelf, they say. Lois has a lot of questions. What kind of genes did they use to improve the tomato? Will it taste any better, or cost more? Will they be labelled so consumers can choose between traditional and transgenic tomatoes? She knows that there are some critics out there who don't think these tomatoes should go on the market so soon, but they don't send press releases and Lois hasn't had time to track them down. She attends the conference but—like the other journalists who are present—doesn't get a chance to ask many questions. Lois shakes the hand of the guy in the life-size tomato suit and accepts a free sample of the new product. She chops it into her evening salad; it doesn't taste weird and Superman likes it. The next day she contemplates ringing those dissidents to find out what they've got against the tomato. Her editor phones and asks how many stories she's got for the next edition. She mentions the possible new angle for the tomato story. But there is a silence: 'Why bother?' he demands, 'You've got a brand new genetically-engineered tomato. It's exciting, it's novel—and why do you want to go upsetting those guys down at Vegtech anyway? We get a lot of stories from them'. She starts writing.

This Lois Lane lives in the real world of deadlines and politics. She is a different Lois from the one writing for the fictitious *Daily Planet*, a person who would never run a rampantly positive story about genetically-engineered tomatoes if there were negatives to ferret out. But the Lois of the Superman series never appeared to have deadlines.

She had time to run around uncovering the truth and getting into scrapes (and clinches) with Clark Kent's alter ego. In contrast, the 'real' world of journalism leaves much to be desired.

This chapter looks at the unequal relationship between journalists and the scientific community and the way this imbalance is reflected in mainstream science writing. It sets out to explore: first, what sort of profile biotechnology enjoys in the mainstream media; second, what factors shape that profile; and third, why that profile should matter to us.

THE CONTENT ANALYSIS

The representative mainstream profile of biotechnology examined here was that depicted in the *Sydney Morning Herald* (*SMH*) throughout 1995. A central aim of its content analysis (undertaken to discover trends in the reporting of biotechnology over a full year period) was to explore the image of biotechnology in the print media, especially in the context of a dearth of public debate. To supplement and cross-check a substantial collection of clippings, an *SMH* database search was performed using the following search terms: gene, genes, genetics, biotechnology, genome, and genetic engineering. The sample was not confined to straight news reports; articles were chosen for inclusion in the sample if they contained any discussion of genetic technology. This was a deliberately broad approach, designed to provide an overview of the pervasion of genetic issues into almost every section of the newspaper. 'Biotechnology' and 'genetic technology' were used as blanket terms for activities taking place in the field of genetics. The final yield was 118 articles.

Each article was categorised as positive, negative or neutral according to its tone. Tone was determined on the basis of each article's overall attitude to genetic technology, and the perceived balance—or bias—of the coverage. A positive article may contain references to 'world-firsts', 'landmark studies' or 'revolutionary' techniques. It is characterised by positive quotes like 'a nifty piece of genetic engineering' (Gilchrist 1995), or, 'experts say this is an important step towards boosting the world's food supply' (*SMH* 16 December 1995), or, 'made using the very latest gene manipulation techniques' (Smith 1995). Quotes praising biotechnology are rarely undermined by the inclusion of negative quotes from sceptical sources. Though varied in subject and style, the positive articles are thus categorised because they appear to accept genetic engineering as cutting-edge technology being used for the common good, particularly in medicine. They tend not to raise questions about ethics or efficacy in the long term.

A further division of the positive category includes articles framed in terms of *breakthroughs* and *discoveries* in the field of genetics, which generally push the 'cutting-edge' angle. This is consistent with Dorothy Nelkin's (1987) finding that science is often reported out of context, as a string of isolated discoveries rather than as an evolutionary process of empirical investigation. This style of reporting picks up only the exclamation mark at the end of a scientific sentence and ignores the explanatory material that comes before it.

A negative article often includes questions about ethics and the long-term potential of genetic technology. It may expose the self-interested side of the industry—egos and cash—or it may seek to provide an empathetic alternative to whiz-bang accounts of advances in the field. One father's sympathetic account of life with his Down's syndrome daughter, for example, questions the push to eradicate such conditions, and such children, via molecular research. Negative articles do not promote unquestioning acceptance of the technology, and occasionally they encourage outright rejection of it. Not surprisingly, these articles fail to boost the profile of genetic engineering.

To fall into the neutral category, an article must not swing noticeably towards either the positive or negative ends of the scale. If it is announcing a genetic 'discovery', for instance, it will include quotes from the sceptics. Neutral articles may also trivialise issues, but not necessarily in a negative manner. Examples are found in the way genetic issues are treated in the Stay in Touch section of the *SMH*, which often adopts a humorous approach to news reported elsewhere in the paper.

Finally all articles were further categorised according to their genetic 'story'. Twenty categories were identified from human disease to environment, to genes and ethics, genes and cosmetic problems, to the much less reported stories such as those about genes and real estate, or about genes and leisure.

TRENDS IN GENETICS REPORTING

Over the year, *SMH* stories displayed a decidedly positive attitude towards genetic engineering. Sixty-seven per cent of the articles (79) were positive (of which 25 per cent depicted breakthroughs and discoveries). Only 16.1 per cent were negative (nineteen), and 16.9 per cent were neutral (twenty). A breakdown of specific story categories shows a preoccupation with genetics and human disease, with forty-seven of the articles (39 per cent) focusing on disease conditions and medicine. Positivity was found to reside in promises of cures, new therapies, new drugs and genetic tests for disease conditions.

The emphasis on medical issues in the *SMH* is no surprise in the light of Sharon Dunwoody's US research, and subsequent assertions that making science information marketable to news consumers involves concentrating heavily on health. Interestingly, policy discussions of biotechnology, directed by biotechnicians, have relied on 'mythic' appeals to the desire for a cleaner, healthier, world (Kawar 1989).

Indeed, a safe way to sell a new technology to the public is to appeal to our hopes and fears about disease conditions (even if much of the hope being peddled remains speculative). Yet despite the high focus on disease and health, few articles have focused on these aspects in relation to the safety problems posed by genetic manipulation in the environment (six stories—5.1 per cent). While stories about tentative steps toward 'cures' for myriad ailments fill our papers, some of the most startling and potentially controversial advances are now taking place in the environmental sector. Some of these have the potential to adversely affect our ecosystems (as highlighted by Wills in chapter 5), including those used to produce the food supply.

The only *SMH* reporting that scraped the surface of environmental issues harnessed, as a news 'peg', the escape of the rabbit calicivirus. Without that 'sensational' incident to report, it seems doubtful whether any of the background articles about other disastrous environmental releases involving exotic or foreign organisms of the past would have been written. The problems caused by insisting on a peg for science news have not gone unremarked: 'phenomena not linked to specific events, such as the greenhouse effect or the growth of an American underclass often go unreported or under-reported until an appropriate news peg arrives to supply the needed event that the coverage requires' (Patterson and Wilkins 1991, p. 26). In addition, problematic biotechnology-environment reporting usually starts with a bang and then fizzles out, like the coverage of the US biohazard debate of the 1970s, which dwindled once the controversy in Congress abated: 'In the end the mass media lost interest . . . there was no longer an activity to report' (Kawar 1989, p. 734).

After human disease, the most concentrated story category was coverage of the business and commercial applications of biotechnology. Seventeen stories (14.5 per cent) related to marketing and biotechnology deals, and they were a relatively common feature in the *SMH's* business briefs. Almost all of these stories were positive, particularly where the commercial benefits were viewed as likely to remain in Australia. Genes and human behaviour came in third at nine stories (7.7 per cent), and the personalities and historical/novelty categories were even with six stories each (5.1 per cent). The latter categories may seem unrelated to genetics, but they illustrate the pervasive nature of genetic references, which can be found in such unlikely places as the Good Living and Spectrum sections of the newspaper.

Interestingly, there were only four articles specifically dealing with ethics and genetic engineering (3.4 per cent). Two of those did not even reflect the *SMH* news agenda: they were letters to the editor protesting about the unethical characteristics of a feature on twin studies ('An ethically questionable separation' and 'Elitist view'). The third piece reported objections raised by the Privacy Commissioner about the ethics of testing babies for HIV and genetic disorders without consent ('Proposal for secret AIDS tests attacked'), and the fourth, buried in the *Northern Herald* (an *SMH* supplement with a small circulation serving Sydney's north shore), tackled the 'human dilemmas' posed by the 'genes revolution'. A further two features dealt with genetics and ethics more obliquely. They are 'Down to Basics' and 'Syndrome Stigmata', both first-hand accounts of life with Down's syndrome children.

The results of this study correlate strongly with Susannah Hornig Priest's (1995) analysis of biotechnology coverage in local US newspapers during 1991 and 1992. Hornig Priest adopted a similar approach to dissecting biotechnology news coverage and found (out of 600 stories) an even higher concentration of positive arguments about biotechnology (82 per cent), a higher percentage of stories focusing on economics (48 per cent), and a similarly low representation of environmental risks (7 per cent) and ethics (1 per cent). Even allowing for differences in categorisation of articles and methodology, both studies show a clear trend towards positive stories and a lack of discussion about issues that may be of concern to the public.

Use of headlines

The framing or tone of news stories—often signalled (not always accurately) via headlines—may affect our reception of their content. Twenty-one articles (17.8 per cent) flagged their topic with 'gene' or 'genetic' in the headline. Only three of those articles were negative, and two were neutral. The remaining sixteen were all good-news genetic stories. The only negative story to get close to the front page was headed, 'How the genes revolution poses human dilemmas' (Llewellyn 1995). The neutral stories, 'New study shows families with heart attack gene more at risk' (Sweet 1995), and 'Genetic links with petty crime claimed' (Wilkie 1995) appeared on pages two and ten respectively. In contrast, ten of the sixteen positive stories appeared between pages one and six, and the remaining six between pages seven and seventeen.

Such favourable placement of positive articles in the *SMH* conforms to Bishop's (1974) findings that reassuring headlines were more attractive to readers' and were associated with increased attention and learning of content. Of the 47 medical/health stories in the *SMH* sample, 26 sported reassuring headlines. Although bad news sells, apparently bad news about health is less welcome. This study suggests that biomedical news is mostly packaged for ease of consumption.

Front page stories

Once news is selected for publication it is subject to the rigours and vagaries of newspaper layout. Articles appearing on the front page of a newspaper have obviously been judged the top stories of the day—and as those that will sell papers. The preoccupations of a reading public, at least as they are perceived by editors, become apparent via trends in page one content.

Of all 118 *SMH* articles, fifteen appeared on the front page (12.7 per cent). Of the fifteen front page stories, fourteen were positive and one neutral. Ten of the positive stories dealt with medical issues (good news cure/therapy stories), one with world food supply, two with commercial interests, and one with sheep breeding. The neutral story (Brown 1995) may have included the comments of sceptics, but sported an appealing sci-fi headline suitable for a front page position: 'Lazarus bacteria shows signs of life after 30 million years'.

THE ROLE OF PUBLIC RELATIONS

The total number of stories generated from public announcements and deliberately accessible activities of the scientific fraternity amounted to 51 (43.2 per cent). Twenty-nine of the articles (24.6 per cent) acknowledged either news agencies and journals used to compile the story, or indicated a straight lift from another publication. An additional twelve stories relied on the announced results of research programmes or studies, and another ten stories gained their information from conferences, forums, or scientific 'events'—which the media is invited to cover via news releases.

With regard to the production demands of newspaper journalism and their effect on the tone of science reporting, the use of ready-made material is very tempting for the journalist. As Dunwoody (1993, p. 10) comments:

> Trying to set up an interview with a scientist may be too risky: if the interview falls through, the journalist has no story. Sitting through a two hour symposium at the meeting may be similarly risky: what if the speakers drone on and nothing 'newsworthy' emerges? Instead, the journalist may rely on things such as press conferences or news releases, products designed to deliver 'news' quickly . . . A news release will not only summarise the main point of a scientist's talk but will also deliver a few pithy quotes; a journalist in a hurry can even use the news release verbatim and no one will know.

A common feature of stories reconstructed from public relations sources is the absence of interviews and challenges to claims. Some reports include quotes which are apparently (although not openly acknowledged as such) lifted from reports in science magazines or recorded during public addresses. This indicates that a large percentage of the reporting of genetic issues relies on the scientific community's desire for publicity, solicited via media releases or the publication of their research results. It also highlights the media's reliance on regular events such as congresses and annual conferences to generate news (Tiffen 1989).

Nelkin (1987) has charted the history and use of public relations by scientific researchers and organisations in the US to attract funding, reassure the government, and avoid unwelcome regulation (with regard to Australia, see chapter 2). She actually uses the example of emerging r-DNA technology in the 1970s to illustrate how a strident media campaign successfully marginalised the critics. The r-DNA controversy and the measures taken to kill it are examined in more detail by Goodell (1986), who concludes that science journalism is heavily influenced and shaped by members of the scientific community and their publicity requirements.

Other work in the US (Mazur 1981) has discovered a similar reliance on public relations, prompting one analyst (see Loge 1991) to suggest that science reporting does not represent the full range of views on a given issue because science reporters wishing to maintain access to scientist sources may adopt their sources' 'vision' of science. Australia's Ian Ward wonders whether public relations effectively blocks access to news space, so that dissident voices are not only struggling against the tide of mainstream views, but are also knocked aside by the waves of publicity material rolling out of mainstream organisations with expensive PR programmes:

> . . . the widespread use of public relations may well close off the public sphere. It has made competition for limited news spaces fierce. As a consequence, many groups with limited resources and no access to professional PR have been effectively prevented from presenting their case via the media. There can be no informed rational political debate, of course, if the media—and hence the public—are given limited information organised around a narrow range of concerns (Ward 1995, p. 162).

The issue of PR-generated science news is underscored in Friedman's (1986) analysis of the Three Mile Island accident in the US. As a consultant to the Public's Right to Information Task force in the wake of the nuclear plant accident, Friedman identified a number of factors which contributed to the local community's confusion and ignorance when things went wrong. The factors included the local media's mostly uncritical use of the highly technical (and thus, often inscrutable) weekly media releases from the plant's PR unit, the failure of plant PR staff to effectively inform the local media, and the deadline pressures of production in media. These factors coupled with a dearth of knowledgeable science reporters left little room for in-depth or investigative reporting. Moreover, Friedman laments the passive approach adopted by journalists writing about Three Mile Island *before* the accident occurred, saying they had enough information to detect major problems then, but did not bother to read between the lines of the plant's regular media releases.

The influence of news values is again obvious in the Three Mile Island example. It bears some relation to the calicivirus coverage, which also escalated and became critical once there was an 'event' to justify it. Media coverage of the rabbit virus gained momentum because traditional news values applied to the story: drama, unexpectedness, conflict, personalities and proximity (Tiffen 1989, Fowler 1991). Readers did not need in-depth explanations of the science involved to follow or 'enjoy' the stories.

The convenience of PR material pitched as hard news, coupled with deadline demands, can thus discourage journalists from digging beneath the surface of a story or producing longer pieces suitable for the features section of the newspaper. Analysis of the 118 *SMH* articles finds that 88 stories (nearly 75 per cent) fall into the hard news or straight report category, 26 (22 per cent) stories are interpretive or attempt to provide background to an issue, and only four (3.4 per cent) represent generalised features or opinion.

NEWS PRODUCTION: THE PRESSURES ON SCIENCE JOURNALISTS

For the average *SMH* reader in 1995, it would appear that biotechnology is advancing bravely into the future and shaping up into a lucrative industry, filled with therapeutic benefits for society. It would not appear to be a field fraught with ethical dilemmas, risky procedures, and, if left unchecked, a field with the potential to develop in undesirable directions.

The overwhelmingly positive, or at least unquestioning, tone of most coverage points to a general acceptance of biotechnology amongst science reporters, or perhaps a reluctance to rock the boat with investigative and critical stories. Much has been written about the relationship between scientists and the mass media, and such work provides clues as to why there is such passivity in science reporting. Goodell (1986), for instance, has noted that the press, like the public, can be intimidated by science, so editors will require their reporters to use credible (prominent and mainstream) scientific sources.

Can we attribute a lack of criticism in science reporting to a lack of specific knowledge on the part of science reporters? Dunwoody (1993, p. 8) finds that US journalists 'typically have little training in mathematics or science' and that this is one of the factors which may influence their acceptance of claims by scientists. A recent Australian study (Henningham 1995) found that although some 75 per cent of 140 specialist science journalists in Australia possessed tertiary qualifications, only 12 per cent held a science degree. Henningham however does not apparently see this lack of specialisation as a barrier to good science reporting, likening Australian science journalists to an elite:

> . . . scientists interacting with specialist science journalists should feel reassured that they are not generally dealing with ignorant hacks who will shamelessly distort their research for the sake of an attention-grabbing headline, but with professional peers who should warrant their trust and co-operation (Henningham 1995, p. 94).

Australian science journalists who wish to remain worthy of the scientific community's time are perhaps willing to embrace the ideologies of mainstream science: chiefly, that science and technology are progressive and for the common good. It is possible to surmise then, that reports will not 'err' on the side of criticism. Dunwoody (1993) suggests that the high status afforded science affects the balance of power between scientists and science reporters; producing a delicate relationship described earlier by Nelkin (1987) in terms of deference and reverence:

> The high status of science, in other words, makes criticism of it risky, particularly for non-scientists. Journalists who cover science on a regular basis are encouraged to buy into scientific truth and to pass on the versions of reality held dear by the powerful scientists of the day. Contrary behaviour is punished by a type of professional snubbing and denigration that can be costly to a science journalist, while compliance opens the doors to Nobel Prize winners. The result, of course, is that the press conveys images that are consistent with mainstream scientific views and that eschew fringe perspectives (Dunwoody 1993, p. 20).

Several studies point to the fact that it is difficult enough for science journalists to forge a fruitful relationship with scientist sources, let alone find the confidence to criticise what those sources have to say (Dunwoody and Friedman 1986; Nelkin 1987; Loge 1991; Dunwoody 1993). This sort of imbalance cannot fail to have an effect on the tenor of day-to-day science reporting:

> In general, science is reconstructed by the mass media as a positive force for society. You will rarely read about science's failures, about the foibles of the scientific culture. Mainstream journalism will rarely criticise mainstream science because they will be punished rather than rewarded for such behaviour (Dunwoody 1993, p. 42).

AGENDA SETTING

The final consideration—the effect of science reporting on the reading public—is harder to assess. Dunwoody categorises and discusses likely media effects thus: media stories as *signals*, and mass media as *legitimisers* and as *teachers*. Media signals are described as the precursors to agenda-setting:

> Each day's media diet produces a set of blips on our environmental scanner. Some of the blips have little meaning to us and will vanish quickly; others may loom as topics or issues to which we may direct some of our attention. When the latter phenomenon takes place, social scientists argue that the media have 'set our agenda' (Dunwoody 1993, p. 28).

Ian Ward (1995) has explored agenda-setting theory in its different guises, summing up the influence of news coverage as a 'cumulative effect' which encourages people to think *about* certain issues without actually telling them *what* to think about those issues. Agenda-setting theory postulates that the issues 'prioritised' via media visibility are seen as more important: 'The essential claim which agenda-setting theorists make is that the news media's own agenda will signal to audiences who read and watch the news that some policy issues are far more important than others' (Ward 1995, p. 50).

In the case of genetics, the regularity, patterns, and tone of the coverage measured in the *SMH* content analysis would seem to indicate that genetic news, especially *good* genetic news, is high on the agenda (that is, highly newsworthy). It even creeps into the nether regions of the newspaper (for example, Good Living where traditional 'hard' news values are less important). When media signals are not quickly forgotten, they have found a place in the news agenda, and genetics certainly fits this category.

The mass media certainly do not function as an effective *teacher* where biotechnology is concerned because of the brevity of reports, the lack of alternative information, and the exclamatory approach to science stories which favours 'breakthroughs' and controversy over routine science:

> This serves to decrease informed discussions of scientific issues, because most science is not of the breakthrough or controversial variety. Instead,

> most science is complex and dull—two things arduously avoided by the popular press. When issues do get discussed, their framing means that lay viewers will consider each event as a breakthrough rather than as a unit in an ongoing research project. This framing plays an important role in opinion formation, and attitudes towards issues of biotechnology (Loge 1991, pp. 6–7).

The mass media seemingly act as *legitimisers* in the case of genetics because they stick to mainstream opinions. They do not adequately cover the 'fringe' views which question the field, and thereby deny critics the 'social sanction' afforded positive accounts of biotechnology. Dunwoody however makes no solid claims about legitimisation: 'No one, to my knowledge, has explored this question empirically in general audiences' (Dunwoody 1993, p. 28). Yet, she does acknowledge that at the basic news-gathering level the mass media treat some views as more legitimate than others. The public, in turn, consumes the 'chosen' perspectives—those considered newsworthy by journalists and editors, who, in turn, are themselves influenced by their scientific sources:

> . . . one attribute shared by the bulk of the issues that do make it onto our personal lists of 'things to be concerned about' is their presence in the mass media. We—scientists included—seem remarkably willing to buy into the notion that what's in the pages of a newspaper or in the line-up of a TV news programme matters, that these few topics are somehow more important than the plethora that doesn't garner such visibility (Dunwoody 1993, p. 41).

If the science news offered is mostly the reconstructed material of mainstream scientists, we might speculate that mainstream views are certainly granted surface legitimisation through sheer visibility. How that visibility may influence the process by which a reader catalogues and legitimises information in his or her own mind is a question that has yet to be satisfactorily answered, and deserves more attention. It seems obvious that the gaps in our awareness caused by the absence of 'fringe' views must affect our attitude to, understanding of, and acceptance of, science information. We are more likely to believe a technology is safe or in our best interests if we have read little or nothing to contradict that idea.

Loge adds to this argument with the idea that 'context' adds to the shaping of understanding: 'The relative importance that the general public places on an issue is related to the relative importance that the media place on an issue. Additionally, the context in which the media place the event is likely to be the context in which the general public places the event' (Loge 1991, p. 6). The Three Mile Island debacle (referred to above) points to the false sense of security that the media can unwittingly engender in a reading public by sticking to the 'safe' contextual environment of media releases.

Finally, this argument is not about media conspiracies (although the 1970s r-DNA controversy should have taught us something), but about highlighting the absence of a comprehensive debate about genetic engineering issues in the mainstream press. It is clear that scientists who talk to the media have their own agenda: legitimisation within the scientific community and amongst the public, leading to funding and support for research programmes. Scientists' willingness to talk to journalists rarely appears to be motivated by a simple desire to inform the public for the good of society in general. It seems equally clear that the demands of production, the adherence to traditional news values, and a heavy dependence on the goodwill of scientific sources, limit the scope of science journalism and erode its potential to be of genuine use to the public.

CONCLUSION

Information—whether scientific or otherwise—does not come without baggage of some kind. The public needs to be aware that the science they read about in the papers is predominantly the science that someone—be it private corporations or public organisations—wants to sell, and that journalists cannot always be relied upon to seek out the whole story. As Ward (1995, p. 164) suggests 'public relations practitioners well understand that, despite the myths surrounding it, the news is not objective. It is not written from nobody's point of view, but constructed from information supplied by informants or sources'.

Many of us would like to believe in the terrier-like approach to reporting we see in mythical journos like Lois Lane, but the reality is that Lois, with her dogged determination to uncover the truth, is not the kind of reporter mainstream scientists invite to poke around their labs. She is usually prevented from getting close to the test tubes, petri-dishes and recombinant-DNA experiments which form part of the new age of genetics. Indeed, the Lois myth needs to be exploded even further, because the modern journalist does a lot less poking around than the romanticised image of Lois on the page and screen—unless you count sifting through piles of press releases.

Hornig Priest (1995) argues that it is in the interests of all parties concerned—the biotechnology community, the general scientific community, and citizens striving for democracy in a high-tech world—to encourage broad discussion of all the issues and view points, negative and positive, that affect public understanding of risk. She also suggests that the media could offer us more in this area: 'information equity in mass media coverage of science—rather than debate dominated by narrow institutional views—should be seen as a journalistic standard' (Hornig Priest 1995, p. 53).

In the hands of policy makers, biotechnology does have the power to affect our lives in profound ways. Unfortunately, a political understanding of related issues (or side-effects) does not always emerge in step with new technology. At some stage we will all come into contact with some aspect of the new biology, making it likely that there will be greater politicisation, expressed in the mass media, of the issues involved. At the moment, political debate at the lay level is limping rather haphazardly behind the rapid implementation of the technology.

I feel sure the fearless and truth-seeking Lois would be disappointed, and with the tenacity that hooked Superman, would jam her spike heel in the door of the tomato lab, track down the gagged dissidents and, armed with all sides of the story, harangue her editor into putting the transgenic-tomato debate on the front page!

3

Bioscience in action! Subduing dissent, containing debate

RICHARD HINDMARSH

'Uproar over mutant meat' *The Age* headline rang out (O'Neill 1990a). The cause for 'uproar' though had occurred two years before its reportage! So-called 'mutant' meat from some 50 transgenic pigs had been sold on the Adelaide market by Metrotec—a joint venture between Metro Meats (Adsteam) and Bresatec (University of Adelaide)—without full authorisation and without public knowledge. A subsequent article headlined—'The issue is the right to know' (O'Neill 1990b)—had identified a key, and now growing, community issue concerning genetic engineering. (Another one here, of course, was animal welfare.)

'Metrotec failed in its duty to put the proposal to us' confirmed Professor Nancy Millis (O'Neill 1990a), head of the federal government's supervisory body over genetics experimentation, the Genetic Manipulation Advisory Committee (GMAC). Yet despite the major breach of the regulatory guidelines, and ensuing strong calls by community organisations and Democrat Senators for an inquiry to be held and for federal laws for r-DNA work, the government-funded research project was not penalised. Funding could have been withdrawn, but it was not. This passive outcome could be considered all the more serious because the University of Adelaide was implicated in another (and earlier) attempted transgenics 'cover-up' from 1984–1986 where those responsible for the breach of regulatory guidelines (in 1984) had also not been penalised upon being 'found out'. Instead, a rather self-serving slap over the wrist had emanated from regulatory head Millis. Her letter of censure to the University's Vice-Chancellor read in part:

> It is most regrettable that eminent researchers have violated the voluntary monitoring system . . . There is genuine concern among the public and within the general scientific community over some aspects of recombinant-DNA research. This monitoring system was originally set up by researchers to regulate themselves in a manner which is responsible, generally effective and *minimises interference in research* [my emphasis].

> However, any violation is likely to cause considerable disquiet in many quarters . . . If incidents like this one were to become common then government may have to revise its attitude towards a voluntary system of monitoring (Millis 1986, pp. 1–2).

The failure to more heavily penalise the errant r-DNA researchers in 1986 had arguably encouraged the later 'mutant meat' incident to occur. Such implied regulatory failure highlights yet another important issue—in-house or peer-review regulatory control of a technology noted by one of its own senior practitioners to be 'more dangerous than splitting the atom' (see ACF 1992). Public trust, actual use of the technology, safety, adequate scope and assessment of the risks, and honest and open accountability, are all at issue here.

A few years later, in 1994, with those regulatory breaches and their associated issues seemingly long forgotten, a *Canberra Times* headline barracked, 'Yes, we'll eat those tomatoes' (Mussared 1994). A pilot survey on public attitudes to genetic engineering had found, 'Most Canberrans would like to try eating genetically engineered tomatoes, and many would willingly grow them in their home gardens'; or so claimed the survey's sponsor, the Department of Industry, Science and Technology (DIST)—a key federal government agency supporting biotechnology as the way forward to a new industrial millennium.

Yet, despite the confidence projected about the survey's pro-biotechnology claims, DIST declined to release its full findings to the *Canberra Times* (Mussared 1994) or to bio-critical campaign group the Australian GeneEthics Network. Later, the main survey would show up highly questionable methodology, and significant consumer concerns, that contradicted the department's claims (see Hindmarsh *et al.* 1995; Hindmarsh 1996). Misinformation or disinformation, especially that disseminated by public servants, is yet another key issue for the public to consider in the debate over genetic engineering.

The above events and the issues raised signal an intense—though little known—debate in Australia now occurring for over two decades. Indeed, within scientific circles it began in 1968. With molecular biologists 'cracking the DNA code', Sir Macfarlane Burnet—Australia's greatest virus researcher—'warned that tinkering with the genes of bacteria and viruses was a dangerous pastime . . .' (Hoad 1997, p. 40).

Since the late 1980s, with field trials of novel organisms underway to develop commercial products, this debate has widened with environmental and consumer organisations taking keen interest. Yet, although these groups have since raised numerous ecological and social issues and important ethical questions (see chapter 1 for example)—biotechnology proponents have successfully inhibited adequate, and sometimes any, consideration of these issues and questions. They have also inhibited open public knowledge about them, subsequent public debate, and attempts to gain stricter regulation or public control of genetic engineering. Indeed, the evidence points to a deliberate attempt to keep the public in the dark for as long as it takes to bring the technology into production.

So how are stricter control of recombinant-DNA regulation and public knowledge and debate being stymied? Why are government agencies conducting politically-sanctioned campaigns to educate the public to the 'benefits' of this potentially dangerous technology without adequate consideration or open community debate about the costs and issues?

These questions are partially answered in this chapter by considering how the social agenda behind the development and regulation of genetic engineering has been constructed. The analysis makes it quite clear that science and technology does not develop in a political and economic vacuum as a value-free, objective undertaking as science would like us to believe. Instead, it is embedded in existing economic and political relations—or, in a sociological sense—social power relations. Here, *power* is exerted by actors through influence and strategy to secure favourable outcomes. As French sociologist Bruno Latour (1987) outlines in his book *Science in Action*, 'Technoscience is war conducted by much the same means. Its object is domination and its methods involve the mobilization of allies, their multiplication and their drilling, their strategic and forceful juxtaposition to the enemy' (Shapin 1988, p. 534). In this 'war of conquest', 'actors work out their impulses to grow, to transform themselves from "micro-actors" to "macro-actors" by subduing others . . .' (Shapin 1988, p. 534).

Two key 'fronts' for bio-proponents—indicated above—in the technoscience 'war' to attain a biotechnological society are those of regulation control and information control. Through in-house regulatory

control, proponents 'organise off' the regulatory agenda consideration of many ethical and ecological issues associated with gene-technology proposals, as well as interrelated social ones like the consequences of the technology's application on people's living and working conditions, or the implications of the private ownership of Nature through patents (see chapter 14). In other words, controlling the regulatory agenda to address only technical-genetics aspects of genetic engineering amounts to the suppression of other equally important, and many would argue more important, issues. Schattschneider (1960, p. 71) terms this the 'mobilisation of bias'. Mobilising bias to predominantly a genetics regulatory basis makes it so much easier for geno-fixing to proceed.[1]

Through information control, images of genetic engineering are constructed that project sanitised and favourable images of the technology. The overall aim is to indoctrinate or manipulate society to accept—

or to be indifferent to—what the bio-Utopians are up to. Herman and Chomsky (1994, p. xi) refer to this process as 'the manufacturing of consent' or as 'the creation of necessary illusions'. In his study of Australian propaganda, Alex Carey refers to it as 'setting the terms of debate', 'managing public opinion', 'taking the political risk out of democracy', or as 'protecting corporate power against democracy' (see Carey 1995).

The capability of proponents to undertake regulatory and information control and to secure enormous research and development funding is of course strengthened by their location in existing structures of domination in the bio-policy terrain—including: the scientific establishment—represented here, for example, by the Australian Academy of Science, the Commonwealth Scientific and Industrial Research Organisation (CSIRO); industry bodies like the corporate-linked Australian Food Council and the Australian Biotechnology Association; and, government agencies that overly support technological change linked to a narrow agenda of economic growth and capital accumulation.

We thus delve into and behind the technocratic decision-making arena that represents the 'invisible' side of government science and technology policy making, and look at how dominant actors apply power to both contain public debate and subdue those challenging bio-Utopia. This illustrates how control is being accomplished and by whom. In turn, this helps to explain the apparent inertia of government to address public concerns raised by any valiant attempts of the media—like that of *The Age* (above)—to get public debate going. In the end, we find that entrenched forces—who use the media to relay a favourable image of biotechnology (see Kawar 1989; chapter 2 this volume)—have established and use an array of strategic channels by which to outflank, and thus contain and subdue, open debate and community concerns. Our account of *bioscience in action* begins at the regulation 'front'.

SETTING THE REGULATORY AGENDA

Following Boyer and Cohen's remarkable recombination of toad and bacterial DNA in 1973 (as remarked upon in chapter 1), some molecular biologists and US environmental groups became agitated about potential (catastrophic) biohazards arising from the shifting of genes across species barriers. Noted US biochemist Paul Berg, prompted by reaction to his own research, also had reservations. Berg was experimenting with splicing the tumour-producing virus SV40 into *Escherichia coli* (*E.coli*), a bacterium commonly found in the human gut, and, according to the *National Times* (Horin 1976, p. 17), was receiving daily phone calls from scientists. 'They'd ask "Send us pSCIOI [a variety of DNA]". We'd say

"What do you want to do?" And we'd get a description of some kind of horror experiment and you'd ask the person whether in fact he'd [sic] thought about it and you would find that he hadn't really thought about it at all.'

The resultant threat of widespread public controversy, led, in February 1975, to the US 'Who's Who' of molecular biology issuing a 'call to arms' to colleagues and allies to air a proposed voluntary moratorium on hazardous r-DNA experimentation. The 'Asilomar' conference brought together 140 leading molecular geneticists, microbiologists and biochemists from around the world, particularly 'veterans' from the national academies of science of industrialised countries, including two sent by the Australian Academy of Science. Although the explicit aim of this conference was to evaluate biohazards and develop safety guidelines, implicit was the consideration of only laboratory experimentation biohazards (excluded were gene therapy or military application of r-DNA techniques, and the broader social and ethical issues).

The strategy was to reassure the scientific community and the general public that gene-splicing could be done safely (see Krimsky 1982) (or at least, genetically safely). A two-fold 'pincer-movement' was quickly deployed by the bio-elite. First, internal scientific disagreement about the narrow scope of biohazards considered at the conference—and the shallow depth of their assessment—was silenced by the persuasive suggestion that any dissidence might threaten research altogether. Scientists were asked to compromise, and to ensure the consensus needed to project an authoritative message to establish control over regulatory policy . To justify such unscientific practice, the scientists reassured themselves that the risks of their work were remote, and were vastly outweighed by the scientific and medical benefits that would flow from their research. In following this path, the scientists congratulated themselves that they could make the necessary ethical and social judgements about r-DNA experimentation; all that was needed to ensure uninterrupted research was to construct an 'acceptable' (minimal) level of safety, sufficient to ward off external scrutiny and the threat of government passing restrictive legislation.

The second tactic in this pincer-movement was to interpret the Asilomar 'review' of the biohazards to the wider community as social responsibility and thereby prove that the scientists could self-regulate themselves voluntarily. Their message became, 'The technology is safe, we are responsible and wise, and we need open flexibility to experiment and develop'. Compulsory regulation of r-DNA work was presented as an unnecessary waste of time and resources, and any ideas of a moratorium were cast aside.

On this basis, the US National Institutes of Health (NIH) (the foremost US biomedical public sector research institution, housed in the Federal Public Health Service) established standards of practice (guidelines) for contained experimentation. To carry these out, in-house scientist oversight committees were operationalised. This became a tactical manoeuvre internationally. Simultaneously, a cohesive network of gene-altering proponents—the bio-policy network—mobilised to strengthen their influence over future policy directions. The regulatory foundations of bio-Utopia were being laid.

In Australia, the Academy of Science Committee on Recombinant-DNA Molecules (ASCORD) emerged. Australia's then science policy was embedded in a non-interventionist market approach. This not only allowed scientists autonomy to pursue their own research programmes, but also presented ASCORD with the opportunity to shape r-DNA monitoring processes in Australia from the ground-up. An immediate manoeuvre was to restrict ASCORD's membership to Academy fellows, allied CSIRO scientists, and colleagues. With the support of government funding bodies for university and public-sector linked research proposals, ASCORD gained ascendancy to allow r-DNA experimentation to occur along the politicised and limited safety lines established at Asilomar.

OUTFLANKING COMMUNITY CONTROL

Despite such manoeuvres to contain controversy arising in Australia, public concerns soon emerged. An unconvinced Sir Mark Oliphant—a noted physicist and Governor of South Australia—warned that genetic scientists were waiting to start research that 'could produce uncontrollable epidemics', and that there was an increasing call to ban such experiments (*The Australian* 1976). In 1977, an ABC *Four Corners* programme and the print media probed further into the international controversy about safety as well as into Australia's measures to ensure safety.

In the Senate (with prior advice from ASCORD personnel), Science Minister Webster issued forth, 'there is no risk to public health in those experiments presently carried out in relation to recombinant DNA research in Australia' (Hansard 1977, p. 1). Department of Health (DoH) bureaucrats apparently thought otherwise. Concerned about a possible r-DNA incident occurring, Department of Health officials thought that should one occur then government could better handle any repercussions than could ASCORD. Because of interlinked interests, the two parties soon reached a mutually-supportive position; the health department and ASCORD both planned to submit similar submissions to Cabinet recommending that r-DNA supervision should shift to a government-convened committee based upon a continued reliance on

ASCORD personnel and procedures, but that also allowed for government monitoring—most, of course, sympathetic.

Yet, as in the scientific community, internal dissent existed within the bureaucracy. For example, Sir Hugh Ennor (the then Science Department Secretary), in reviewing ASCORD's proposed submission, wrote to the health department counterpart Cyril Evans: 'I note with some disappointment that the Academy has not commented on the desirability of stimulating public debate . . .' (Ennor 1977, p. 1). Evans (1977, p. 1–2) replied,

> It was my impression that your Minister concurred with the view that a formal public inquiry was not warranted. It seems to me that he took the initiative in arranging for an informal presentation of the issues by Professor Ada to Parliamentarians as a substitute for this and future debates . . . I understand that the reactions of the Parliamentarians were quite moderate. I believe that there are dangers in fostering additional public debate . . . I, personally, consider that better decisions will be reached if the matter is not carried to the extremes reached in the USA.

In his talk though, Ada—the ASCORD chairman—had played down the issues and presented genetic engineering as quite a normal activity that was just a 'little different'. Soon after this, other science bureaucrats (from the policy division) played down Ennor's views to a new Science Secretary by pointing out a general easing of the public controversy overseas.[2]

That 'general easing' itself however was the outcome of an intense mobilisation of the international bio-elite to defend the self-regulatory front still under residual but entrenched blockade in the US from the safety debate that would-just-not-go-away. It had not been subdued easily as the scientists desired. Mounting a defiant rearguard defence, a series of top-level international meetings—between 1976 and 1978—had been organised by the NIH and the European Molecular Biology Organisation. These unannounced and private meetings were restricted to a select 'phalanx' of scientists. Following a familiar pattern, any internal dissent and contrary views to the bio-elite were organised off the agenda, as the elite enveloped these meetings in a 'state of siege' atmosphere: us against them; science against 'antiscience'; proponents against critics; biodevelopment against research annihilation; and so forth (see Wright 1986).

As had occurred at Asilomar, the agenda became further politicised by introducing restrictive assumptions concerning *degrees* of safety rather than a commitment to safety *per se*, which included limiting analysis to a subset of biohazards. The central assumption made was that *all* r-DNA research was conducted with *E.coli* strain K12 (even though NIH guidelines catered for other host organisms). Important issues about

E.coli K12, such as low-level seepage of novel gene combinations from research sites into the environment, were submerged, and the hypothesis that the scientists supported was 'that whatever else might be done to it, it was impossible to convert *E.coli* K12 into an epidemic pathogen which could escape the laboratory and run rampant through a population'. This became known as the 'epidemic pathogen hypothesis' (see Wright 1986, p. 598).

Back in Australia, r-DNA operatives became increasingly blatant in their bias. At Australia's first general scientific meeting considering r-DNA technology in 1977, internal scientific dissent again surfaced when Professor John Walsh—a University of NSW geneticist and Dean of the Faculty of Medicine—suggested that precautionary action in the form of external legislation should be considered. His suggestion was met by a 'deafening silence' and further organised off the conference agenda by ASCORD members in attendance who dismissed public misgivings about r-DNA experimentation (Hoad 1977).

At about the same time, the first major skirmish in Australia between community interests and r-DNA practitioners erupted. When the Walter and Eliza Hall Institute of Medical Research began constructing a C3 (containment of high risk) laboratory at Parkville for potentially hazardous r-DNA experiments, the University of Melbourne Assembly launched a high-profile public inquiry. In voicing its protest to the project, the Assembly exposed the logical paradox within the scientists standard safety defence of their research: if experimentation was so safe, then why were such elaborate containment procedures required? After a two-year long inquiry the Assembly's report was published in 1979—incidentally, the same year that the CSIRO declared biotechnology as a priority research area. The report comprehensively criticised Australia's regulatory system and called for an immediate halt to r-DNA research while further evaluation of the risks was undertaken.

The embattled bioscientists subsequently 'battened-down'—with the support of the University of Melbourne—by rejecting this call, as well as others from prominent citizens calling for further explanation. In launching a counter-attack, the epidemic pathogen hypothesis was 'wheeled-out' to vigorously 'engage the enemy'. For the 'front page', a press release stating that genetic engineering was safe was quickly dispersed. At internal meetings with government officials the Assembly report was 'torpedoed'. At one of these meetings, Department of Science and the Environment (DSE) and other bureaucrats sought reassurance about applied genetic engineering from an upper echelon of scientists from the Australian Academy of Science and the CSIRO (Farrands 1979). Desiring to retain self-regulation, and advocating no public debate, the scientists reported that an Academy report reviewing r-DNA technology was ready for a public dissemination exercise. Enrolled to this strategy, the bureaucrats hastily requested a promotion of the technology's future commercial benefit as well.

Predictably, the reassuring report—'Recombinant-DNA: An Australian Perspective' (Australian Academy of Science 1980)—promoted the 'authoritative' 'battle-cry' of the epidemic pathogen hypothesis, emphasised that r-DNA work was safe, and promoted its commercial opportunity. To handle industrial application and possible environmental release of new organisms, it recommended that a surveillance committee be set up by government with a continued reliance upon ASCORD personnel and procedures. DSE Minister Thomson rubber-stamped the (pre-arranged) recommendation, and the new RDMC (Recombinant DNA Monitoring Committee) was established in late 1981 within the *safe* territory of the pro-biotechnology Department of Science and Technology (DST, formerly DSE). The report's press release, in turn, was relayed to the public through favourable print media coverage, like that of *The Australian* which stated the report 'should go a long way towards dispelling "old wives tales" about genetic manipulation' (Brumfield 1980, p. 9).

The skirmish to assert some community control over r-DNA regulation was thus short-lived, and the principle of self-regulation firmly established. Australian scientists (in alliance with their colleagues overseas) then began to promote highly speculative claims about biotechnology's revolutionary scientific and commercial capacity, attempting to bury once and for all concerns about both the safety and desirability of r-DNA technology, as well as to attract the enormous funding resources needed for the enterprise of reconstructing Nature.

Internationally, economic opportunities, industrial competitiveness and scientific leadership became the driving agenda of biotechnology, its future linked to national prestige and wealth. Ethical and environmental concerns became buried in the 'real world' of economic development, trade and high finance. Revolutionary industrial science, academia, politics and capital fused in a heady mix at the global bio-business frontier of the US. Commercialisation began to significantly influence the policy agenda globally.

COMMERCIAL OPPORTUNISM

For emerging global bio-entrepreneurial conquest, Australia hastily raised its banner. In 1981, the DST's annual report emphasised the new mood of international competitiveness by referring to biotechnology. A *Financial Review* article 'Biotechnology fails to interest Australian industry' (July 1981) became the first of many 'biotechnology: Australia is missing out' type articles (cited in Howarth 1984, p. 40). CSIRO (1981) contributed to the hype of the biotechnology promise, and the Science Minister enthused to one and all that Australia was in the world's 'top ten' in biotechnology research (Eckersly 1981, p. 7).

Commercial interests responded. In 1981, Australasian firm Fielder Gillespie invested $1 million into monoclonal antibodies research at the Queensland Institute of Medical Research. Government infrastructure programmes began to emerge. More Australian universities turned to r-DNA experimentation. By the end of 1982, Austgen, CRA Australia, and BHP were collaborating with university laboratories undertaking r-DNA research.

Yet, hampering bioscientific progress were powerful reservations held in the bureaucracy about loosening Treasury and Finance purse-strings to support such interventionist (as opposed to free market) science policy (see Joseph and Johnston 1985). Subsequently, the issue of high-technology development became elevated to an electoral one by the Labor Opposition in the 1983 federal election battle. (Since that election—won by Labor—public monies have been used liberally by Australia's federal and State governments to implement many measures supportive of high-technology and biotechnology.)

REGULATORY SKIRMISHES RENEWED

But behind the glowing election promises of high-technology, Science department officials had again become worried about r-DNA regulation—now located within their own institutional halls. They were particularly concerned about the RDMC's autonomous style of operation in defining its structure and standards, but less so about its narrow scope of risk assessment. In the latter area, DST officers joined with RDMC scientists to contest the directive to the RDMC by the DST Minister to appoint animal ethicist Professor Peter Singer. The bio-elite won by claiming that RDMC terms of reference emphasised only a technical definition of safety (Fuller 1983). The RDMC, of course, was only too aware that sections of the public remained extremely apprehensive, especially on moral or ethical grounds, about r-DNA work (RDMC 1982, pp. 3–4).

Indeed, r-DNA work at that time—and its limited scope of regulatory assessment—was becoming the subject of increased public scrutiny worldwide. Commercialisation meant new agricultural products in the form of novel organisms being released into the environment. Campaigns against r-DNA technology, particularly in the US, renewed. Serious doubts were raised about the ability of bioscientists—with their narrow reliance on genetic theory—to assess the broader ecological risks. Then, as now, environmental variables were by no means well understood (as Wills emphasises in chapter 5). With the prospect of environmental release, the risks of genetic engineering had dramatically increased (see TWN 1995).

With the international debate growing about environmental release the first proposal for an environmental release in Australia was made. It was to field test a live bacterial vaccine in animals. Either cracking under the intense pressure of the debate, or in exercising good judgement, the RDMC scientists told Barry Jones—the then Minister for Science and Technology—that they lacked expertise to assess the release of transgenic organisms into the environment. Even so, they maintained that the RDMC was 'the appropriate body to assess the hazard associated with *genetic* (my emphasis) aspects of agents proposed for release . . .' (Millis 1984, p. 1). No other option was however suggested to assess the broader *ecological* aspects. Ethical questions also went unaddressed, not only environmental ones but we should remember that animal ethicist Singer had been barred from the RDMC a short time before the proposed release. Then, as a reminder about the social context of environmental release, Sydney University biologist Ditta Bartels argued:

> . . . the recent Australian proposal . . . should have provided a unique opportunity to engage the community in a dialogue concerning the new directions in which recombinant DNA work is proceeding. But instead . . . there has been no disclosure . . . *Commercial-in-Confidence* has been stamped on both the proposal and the documents relating to its assessment . . . all details . . . are kept securely behind the closed doors of the RDMC (Bartels 1984, p. 183).

To make matters even worse for the bio-alliance, now seriously battling to maintain regulatory control on the limited technical grounds of genetics science alone, the Environment Department began launching 'forays' into DST territory to try and 'abduct' the RDMC on conflict of interest grounds. Within the government bureaucracy it was not normal practice for the monitoring or regulation of potentially hazardous activities to lie with agencies responsible for their development. The Environment Department's position was boosted by support from environmental heavyweights, the Australian Environment Council and the Australian Conservation Foundation. All environment bodies wanted more ecological assessment and representation on the RDMC.

With bio-proponents facing loss of, or weakened, regulatory control through 'enemy' outflanking manoeuvres—the bio-policy network strategised that the bioscientists urgently needed to be *seen* to be well in control of the technology, and to be *seen* exercising that control in a *neutral* bureaucratic setting (but well out of range of the 'greenies'), if the genetics scientists were to retain regulatory control.

Following a flurry of counter-manoeuvres, the bio-alliance eventually won that immediate battle through a two-fold adjustment to regulatory policy. First, the RDMC was transferred to the seemingly neutral Department of Administrative Services and renamed the Genetic Manipulation Advisory Committee (GMAC). Second, its scope of experts was expanded slightly to include (limited) ecological input (see Hindmarsh 1994). These moves temporarily 'absorbed' conflict, which was further contained by a consensus view reached within the bureaucracy that it was *not* necessary to initiate public debate about environmental release and genetic engineering.

Undoubtedly, the federal government's prioritising of biotechnology that year (1987) had had a significant influence on events. The *Plant Variety Rights Act* had been passed which further strengthened patent legislation for manipulated lifeforms, and regulatory guidelines were further relaxed (apparently to counter the stock market crash which had resulted in market capitalisation dramatically declining for biodevelopment). Interest in biodevelopment subsequently renewed. Transnational companies like Groupe Limagrain and ICI began to invest heavily, and some 50 Australian biotechnology companies and research institutes earnestly renewed developing collaborative R and D agreements with foreign companies.

In response, public protest slowly mobilised for the next melee. More community groups entered into the conflict zone including the Australian and New Zealand Federation of Animal Societies, Friends of the Earth (Fitzroy), and the Australian Consumers Association. The new Minister with responsibility for GMAC—Stewart West—came under increasing pressure from these groups to hold a government inquiry into genetic engineering. The Australian Conservation Foundation (ACF)—in adopting international resistance manoeuvres—raised those stakes by demanding publicly that a five year moratorium be placed on all deliberate environmental releases of genetically engineered organisms.

PARLIAMENTARY INQUIRY

More pressure for a parliamentary inquiry to be held came from the exposure of Science Minister Barry Jones to graphic images of crippled pigs from r-DNA experimentation on an ABC *Four Corners* programme (11 September 1989). Facing difficult and emotive questioning, Jones promised an inquiry. The final catalyst for an inquiry emanated from the fledgling industry itself, six months after the *Four Corners* programme. Unexpectedly, the Adelaide University–Metrotec 'Case of the super pig that went to market' surfaced (Dredge 1990). Seizing the opportunity, environmental groups bayed: 'An inquiry is needed into the secret releases and the attempted coverup' (ACF 1990). In damage-control mode, bio-proponents GMAC regulatory head Millis and CSIRO Plant Industry chief Jim Peacock argued against restrictive legislation (see Mills 1990). Enough damage had occurred however to breach the walls of self-regulatory control and 'contained' public debate.

In June 1990, the House of Representatives Standing Committee for Industry, Science and Technology received its terms of reference from the industry and technology department's Senator Button to hold an inquiry into genetic engineering. A catch-22 rider existed though—for people making submissions to the inquiry—in its terms of reference. These stated an *a priori* and unqualified acceptance of the 'existing and potential' benefits of r-DNA work. Despite that bias, 35 per cent of the 167 submissions—those of the public—chose to ignore the rider and called for an immediate halt to genetic engineering work in Australia (see Hindmarsh 1994, p. 382). Burying such 'hostile' calls, the inquiry's chair Mr Lee, in tabling its report (Lee 1992a), reinforced the inquiry's bias: 'The Committee believes that the possible economic, environmental and health benefits derived from genetic manipulation techniques are worth pursuing, even if not all of the claimed benefits materialise' (Lee 1992b, p. 1). No mention was made of the significant public dissent to genetic engineering.

Predictably, the inquiry report rejected calls for a moratorium on r-DNA work and instead embraced the industry and scientists' perspective of an internationally competitive regulatory environment. It also intended the compliance of potential 'rogue' researchers to GMAC guidelines by a proposed shift from voluntary compliance to mandatory (though weak) legislation (see Hindmarsh and Hulsman 1992). Not surprisingly, the role for public participation was token. Although the ACF (1989) initially considered the inquiry's instigation to be a 'major victory', that victory turned into a 'massacre' for campaign groups as the inquiry effectively replenished the proponents' arguments and 'authoritatively' 'swept aside' many community concerns. Protest was thus 'absorbed'. Since the inquiry, its report, 'Genetic Manipulation: The Threat or the Glory?', has been used by proponents in various forums as a 'persuasive' vehicle to further shape and construct social reality to the bio-elites' views of the benefits and the 'low risk' of genetic engineering, and to legitimise these views as central in the policy terrain.

Eight months after the report was tabled, the federal government formed a consultative group to negotiate a harmonised State–Commonwealth regulatory framework. Four years later, no harmonisation had been achieved. Indeed, it took until late 1997 for the federal government (in a process facilitated by the industry and technology departments) to just endorse the inquiry's proposed statutory legislation. Another round of State–federal consultation can be expected in 1998, but this time the bureaucrats have included the participation of industry, consumer and environmental groups (Anderson *et al.* 1997). Inclusion of the latter two groups apparently amounted to 'public participation', but as before—in another exercise of the mobilisation of bias—the benefits of r-DNA technology were flagged by government Ministers in charge of the process as a virtual *fait accompli* (Anderson *et al.* 1997).

In the interim, the bioscientist-dominated GMAC has continued to control the voluntary in-house regulatory regime; another three (known and apparently deliberate) breaches of the guidelines have occurred; commercial planting of genetically-engineered cotton has been approved by the National Registration Authority despite many criticisms (see chapter 12); proposed labelling standards for gene-tech foods have been considerably delayed and weakened apparently under intense industry pressure; the number of environmental releases for field experimentation has increased—especially for herbicide-tolerant crops; intellectual property ownership for gene-splicing activities has retained its position as the proponent's top issue of the day; and the federal government allowed the importation into Australia of 'Roundup Ready' soy beans despite the global and local protest to the beans by public interest organisations.

In opposition, public concerns and resistance continue to mount. This is due in part to the on-going campaigning by resistance groups, and in part to the emerging production-line of genetic engineering itself exposing important ethical questions and social and environmental problems. Outstanding examples that have attracted worldwide attention and condemnation from many quarters include 'Dolly' the cloned sheep, the headless frog embryo, the Roundup Ready soy beans (see chapters 1 and 14), non-germinating seeds, and the possibility of human clones.

To counter and contain resistance, proponents—in a continuation of bio-Upotian 'war-games'—are increasingly turning to the tactic of creating a favourable environment for genetic engineering through image-construction. A central part of that image-construction is designing 'public education' programmes to manufacture consent.

MANUFACTURING CONSENT

In 1981, a short time after the 'corpse' of the University of Melbourne Assembly inquiry report had been laid to rest, lingering public disquiet about the possible biohazards of r-DNA experimentation prompted the CSIRO Advisory Council to recommend to the CSIRO Executive that it should 'develop a *public-awareness* [my emphasis] strategy linked, in terms of content and timing, to the announcement of significant CSIRO developments' (CSIRO nd, Appendix B, p. 2).

Less than a decade later, in 1990, a new strategy had been forged. Enrolled to the notion that CSIRO's future depended 'upon the future release and widespread usage of genetically engineered plants, animals and other organisms'—and recognising this as an ongoing 'sensitive' community issue, top CSIRO officials decided upon 'a major, well developed strategy . . . to facilitate the future release and *acceptance* [my emphasis] of this new genetic material' (Reeves 1990, p. 1; see also Peacock 1990; Stocker 1990).

Allies mobilised to support the strategy of public acceptance (though, of course, it is not often called this). In 1990, the Australian Biotechnology Association (ABA)—the genetic engineers' trade association with a membership of business interests, scientists (including CSIRO and university scientists), and government officials—began disseminating widely (without charge) to secondary schools an information pamphlet series presenting an informative account though overly beneficial image of biotechnology.

In 1992, during the latter stages of the parliamentary inquiry, CSIRO took affirmative action. It launched a $250 000 'Genetic Engineering: Will Pigs Fly' exhibition across Australia which clearly sought to enrol public acquiescence to genetic engineering (see Hindmarsh 1992; Love 1993). Its sponsors included CSIRO; the Department of Industry, Technology and Commerce; CSL Ltd; and the cotton industry.

In 1995 another two organisations shored-up the information control front. GMAC's Public Liaison Committee began also to disseminate to high schools a US series of pro-biotechnology brochures. The 'Gene Technology Information Unit' (GTIU) then emerged flamboyantly to lead the charge. With a two-year $500 000 contract from the Department of Industry, Science and Technology (DIST) it surfaced in the terrain to supposedly 'more professionally' create a favourable image of genetic engineering. Its specific target for 'persuasion' was science teachers and their students (the next generation)—the most favoured target of an international campaign to secure public acceptance (OECD 1992, p. 2). The GTIU's flagship—its glossy brochure called 'Gene Technology at Work' (GTIU 1995)—paints a winning picture of the views of biotechnology proponents while trivialising those of critics (see Hindmarsh 1996). Overall, the image projectiled by all these 'public education' programmes is a 'laundered' version of the technology—the benefits are built-up and within that frame the risks and broader problems are submerged and largely sunk as being trivial.

CONCLUSION

Our 'war-correspondent' excursion to the regulation and information control 'fronts' of *bioscience in action* has revealed that, *despite* important community concerns about genetic engineering and increasing calls for open public debate and information, an environment of weak regulations and lack of information has instead been proactively crafted. Consequently, genetic engineering is slowly creeping into our society via a number of routes that are not obvious to the uninformed nor to those kept in the dark. Clearly, in-house regulatory control, biased public education programmes, and contained public debate, are central avenues of manufacturing consent in Australia for bio-Utopia. With its primary agenda of bioscientific–industrial expansion for a new millennium of capital accumulation and corporate power, public concerns and critical views are being effectively marginalised by the political 'macro-actors' of the bio-elite. They reside strategically as a drifting and recombining network of allies—behind the closed doors of stacked regulatory committees, of myopic and self-interested industry associations, and of government bureaucracies unrepresentative of the general public—subduing dissent and containing debate. As a result, the potential spectre of Nature and society being biotechnically-reconstructed is now staring us—the community—in the face.

NOTES

1 Regulation based on this approach also consolidates the profit-making side of genetic engineering. This is based on an industrial-machine worldview of Nature of reducible functional and rearrangeable 'bits and pieces' of genetic information coding for this and that to function. This facilitates those bits and pieces to be fractured off from the whole and to be patented. Yet, despite keen scientific opposition to the 'Master Gene' viewpoint (or Central Dogma of molecular biology—see chapter 5, also Hubbard 1995), it predominates in both the scientific community and in the public mind. Again this relates to the suppression of debate and thus also amounts to information control.

2 Indeed, the recommendation was made that—in the joint DoH/DoS (Department of Science) submission to Cabinet—the 'submission merely note the possibility of criticism from the general public, and *seek to forestall such criticism* [my emphasis] by emphasising that the present situation need not be the government's final word on the issue' (Goleby 1978, p. 4).

4
Enclosing the biodiversity commons: bioprospecting or biopiracy?

JEAN CHRISTIE

> Australia is the only developed megadiverse country which has a strong science and technology capability. However, despite this capability, Australia does not access or use its genetic resources to any great extent. Instead, Australia has tended to access and use exotic genetic resources . . .
>
> (ARC 1994)

> The Western system is seeking to replace or destroy the communal regimes of intellectual property rights. It is the pharmaceuticals industry which stands to gain most from indigenous knowledge regarding biodiversity. From the earliest explorers to the present, Australia's native plants have been subject to investigation for their commercial potential.
>
> (HENRIETTA FOURMILE, BUKAL CONSULTANCY, CAIRNS, 1995)

Australia occupies a unique place in the 'North–South' debate about biodiversity and who can use it. Though sitting squarely with the cash-poor biodiversity-rich countries of Africa, Asia and Latin America that comprise the biodiverse 'South' geographically, Australia aspires to play in the big league of corporate biotechnology with the cash-rich, biodiversity-poor, industrialised 'North'.[1] In the quest for industrially useful biodiversity, Australia has cast its lot firmly with the plunderers and resource users of the industrialised North, in a growing political debate about who shall have access to and control over the earth's biological resources, and on what terms. Yet, Australia is more likely to be plundered than be a plunderer, more likely to be resource provider than resource user.

Australia is already a target in industry's search for commercially useful biological resources. Corporate bioprospectors from Japan, Europe and the USA have set their sights on Australia's biological bounty. Gambling that its own biotechnology capacity will someday grow to compete with these transnational giants, Australia has aggressively adopted the stance of the industrialised North in international negotiations about 'access to biodiversity' and intellectual property over living organisms. Setting itself up as a junior partner to industrial giants, Australia has opened itself to corporate commercial exploitation, ignored the ethical questions associated with 'life patenting' in the biotechnology industry (see, for example, The CornerHouse 1997), and offered scant protection to Aboriginal people and Torres Strait Islanders, whose knowledge of biodiversity will increasingly become the target of bioprospectors from within Australia and around the world.

Biotechnology is a multi-billion dollar industry, and expanding daily (Rural Advancement Foundation International 1996a). Yet despite a growing armory of sophisticated techniques at their disposal, biotechnologists must still rely on the biological wealth of the earth for their 'raw materials'. Animals, plants, microorganisms, and their microscopic parts are—and will remain—the foundation of the life industries. These 'genetic resources' may come from farmers' fields, the fishing and hunting grounds of indigenous peoples, state lands, national parks, deep sea trenches or thermal vents. They may be used essentially as they are found, their proteins isolated or purified, or their genes bioengineered from one species into another. They may be put to one of hundreds of possible industrial purposes. Whatever their source or eventual use, however, living organisms remain the essential life blood of the new biotechnology industries. Without them, the industries themselves would die. This simple fact is at the geopolitical heart of the world's interest in 'biodiversity'. Industry's need to control living organisms fuels a growing controversy over whether living things should be subject to 'intellectual property' claims. And it explains why biodiversity and intellectual property control over living organisms have become the subject of heated debate in two important international agreements signed in the 1990s: the Biodiversity Convention and GATT (the General Agreement on Tariffs and Trade), especially the latter's agreement on Trade Related Aspects of Intellectual Property Rights (TRIPS) now administered by the World Trade Organisation (WTO). Of further impact is the WTO's Agricultural Agreement, and the Multilateral Agreement on Investment, now being discussed. Both erode national sovereignty and extend the rights of transnational industry.

BIOPROSPECTING, BIOPIRACY AND INTELLECTUAL PROPERTY

There is no escaping the fact that the world's biodiversity, however it is measured, is overwhelmingly found in the tropics and sub-tropics of Africa, Asia, Latin America and Australia—thousands of kilometres from the major life industries of North America, Europe and Japan. The South therefore stands to gain, or lose, a great deal in the global race underway to control and profit from the earth's biological resources. The statistics on biodiversity speak volumes. Tropical forests contain at least half of all known plant and animal species worldwide. A 50 hectare area in Malaysia may contain 830 native tree species, while all of Europe north of the Alps has only 50. There may be more plant species in Botswana, Lesotho, Namibia, South Africa and Swaziland than in any other region of comparable size in the world. Biological diversity in the deep ocean rivals that of tropical rainforests, and new research reveals that deep sea species diversity is richest in the tropics, diminishing as one moves toward either pole. Coral reefs are home to more species than any other marine ecosystem. Over half of this diversity is in the western Pacific, and about 15 per cent in the tropical Atlantic. The Indo-Western Pacific has an estimated 1500 fish species, and over 6000 molluscs, compared with only 280 fish and 500 mollusc species in the eastern Atlantic. Tropical Lake Malawi has 245 species of fish; Lake Windermere in the UK has only nine. Brazil has over 3000 freshwater fish species—three times more than any other country in the world. Indonesia ranks first, second, sixth and eighth in the number of endemic (or unique) species of birds, mammals, reptiles and amphibians respectively, while Britain has few or no endemic species in any of those four categories (see Shand 1997, pp. 8–9).

The potential economic value of all this biodiversity is staggering. Not surprisingly, agricultural biodiversity is also far richer in the South than in the temperate North. Every one of the world's main crop and livestock species has its centre of greatest genetic diversity somewhere in Africa, Asia, Latin America or the Middle East, in the areas where they have been longest cultivated. Apart from an emerging 'bush tucker' industry which is beginning to capitalise on the rich food heritage of Australia's indigenous peoples, virtually all of Australia's agricultural commercial production depends on 'exotic' (foreign) genetic stock. To a significant extent, this also is true for all other 'developed' countries. Quite literally, the gene pool that world agriculture must rely on is kept alive and developed by the daily work of small family farmers of the 'developing' countries. Their crop varieties, adapted over centuries to thousands of unique ecological niches, hold the genetic secrets of disease and pest resistance, and adaptability to climate change, that industrial plant breeders and farmers must rely on for years to come. RAFI (1994) has estimated that the annual value added to industrialised country agriculture just by farmers' seed varieties held in the gene banks of international agricultural research centres is US$4–5 billion.

The commercial implications of 'traditional knowledge' and biodiversity are even more dramatic when one considers the world's pharmaceutical trade. For instance, 57 per cent of the 150 most frequently-prescribed brand name drugs in the USA in 1993 were derived from natural sources (Grifo *et al.* 1997), and the percentage is expected to rise with the advent of modern biotechnology. In 1996, global pharmaceutical industry sales were already a whopping US$222 billion.[2] Medicines from natural sources are a growing multi-billion dollar industry.

Western scientists have barely begun to assess the medicinal potential of the world's biological resources. Fewer than 15 per cent of Australia's species, for instance, have even been described by modern science (Glanzig 1996). But the world is rapidly waking up to the fact that most knowledge about the earth's known biodiversity is held in the minds and cultures of the indigenous and rural peoples who have used it for millennia. Nobody is learning this faster than industrial scientists. 'Traditional knowledge', once thought to be irrelevant to the industrialised world, is acquiring a new value—as industry grasps the importance of indigenous peoples' science, and ethnobiologists are hired to scour the globe in search of living organisms whose properties have been known and used for generations. Once these properties are 'discovered' by modern science, and described in Western terms, industry increasingly uses intellectual property laws to appropriate both the biological resources and the knowledge about them as private property. Community innovators who may have created the 'resources', and led scientists to them in the first place, go unacknowledged and unrewarded. By legal sleight of hand, the living organisms themselves are deemed inventions of human intellect, and become the subject of monopoly patent claims (see also chapter 14). And here the fundamental conflict arises, and accusations of 'biopiracy' are made.

In most indigenous and rural societies, knowledge and innovation are not seen as commodities but as community creations, handed on from past to future generations. The earth and nature are used, developed, analysed and managed, but are not exclusively owned. By contrast, European-based intellectual property regimes and the international conventions that have taken their lead, are founded on the belief that innovative ideas and products of human genius should be legally protected as private property. In the last few years, intellectual property has evolved to encompass living organisms. Plant breeder's rights and recent applications of patent law increasingly cover a vast array of living things that are considered products of human genius, and subject to private monopoly controls.

In the 1990s, these conflicts of world view and values have come into sharp focus in the international negotiating arena. Indigenous peoples, farmers' organisations and environmentalists, with some vocal scientists, public sector researchers, ethicists and academics from around the world have challenged industry, and called for 'No Patents on Life', arguing that life forms (including human genetic parts) should not be privatised by intellectual property claims. In the rarefied atmosphere of international conventions, these issues are being hotly debated, because intellectual property is increasingly seen as the newest mechanism for the industrialised countries of the North to gain monopoly control over the resources of the South, with the blessing of international law.

TWO INTERNATIONAL AGREEMENTS: THE BIODIVERSITY CONVENTION AND 'TRIPS'

The International Convention on Biological Diversity (or Biodiversity Convention) was adopted at the Rio de Janeiro 'Earth Summit' in June 1992. It affirms that nations have sovereignty over the biological resources within their borders, but requires that 'intellectual property' over living organisms must be respected. It also recognises the importance of 'traditional knowledge' in the conservation and use of biological diversity. Hidden in these concepts are a multitude of contradictions, biases and inequities (see Table 4.1 below for details relevant to intellectual property and indigenous knowledge).

Six months after the Biodiversity Convention came into force (in December 1993), the 'Uruguay Round' of GATT was signed at Marrakech, Morocco. On 1 January 1995 the World Trade Organisation came into being, to administer and monitor the GATT and the ongoing process of global trade 'harmonisation' which it called for. For the first time in history, a global trade accord contained explicit obligations for signatory states to adopt intellectual property laws, including monopolies over living organisms. This came to be known as the 'TRIPS Agreement'.

Taken together, the Biodiversity Convention and the WTO's TRIPS Agreement demand laws and practices that have profound ethical and economic implications for all governments and peoples. Article 27 of TRIPS, for instance, obliges all signatory states to introduce intellectual property laws over 'life'. Specifically, TRIPS requires WTO members to provide for patents on microorganisms, and to have some form of intellectual property over plants. 'Animals' may be excluded from patentability, though it should be noted that in practice the microscopic parts of animals including humans are treated as microorganisms. Governments' failure to introduce such intellectual property laws could be construed by the WTO as an unfair trade barrier, and the 'offending' country would then be subject to trade sanctions.

Article 16 of the Biodiversity Convention stipulates that Western-style intellectual property regimes must be respected. As governments

Table 4.1 The Biodiversity Convention (intellectual property and indigenous knowledge)

The Convention on Biological Diversity—a multilateral agreement—had been signed and ratified by 169 governments as of mid-1997. These governments (notably excluding the USA, which has not ratified the Convention) are the 'contracting parties' in the text below. The extracts are especially relevant to biodiversity, indigenous peoples' knowledge, and intellectual property rights. Because the Convention is an agreement among states, any protection offered to indigenous and rural communities is therefore via sovereign states.

Preamble, point 12: [Recognises] the close and traditional dependence of many indigenous and local communities embodying traditional lifestyles on biological resources, and the desirability of sharing equitably benefits arising from the use of traditional knowledge, innovations and practices relevant to the conservation of biological diversity and the sustainable use of its components.

Article 1, Objectives: The objectives of this convention . . . are the conservation of biological diversity, the sustainable use of its components and the fair and equitable sharing of benefits arising out of the utilisation of genetic resources, including by appropriate access to genetic resources and by appropriate transfer of relevant technologies, taking into account all rights over those resources and to technologies.

Article 2, Use of Terms, point 13: '*In situ* conservation' means the conservation of ecosystems and natural habitats and the maintenance and recovery of viable populations of species in their natural surroundings and, in the case of domesticated or cultivated species, in the surroundings where they have developed their distinctive properties.

Article 3, Principle: States have . . . the sovereign right to exploit their own resources pursuant to their own environmental policies, and to ensure that activities within their jurisdiction or control do not cause damage to the environment of other States . . .

Article 8*, In-situ *Conservation, clause (j): Each Contracting Party shall . . . (j) subject to its national legislation, respect, preserve and maintain knowledge, innovations and practices of indigenous and local communities embodying traditional lifestyles relevant for the conservation and sustainable use of biological diversity and promote their wider application with the approval and involvement of the holders of such knowledge, innovations and practices and encourage the equitable sharing of the benefits arising from the utilisation of such knowledge, innovations and practices.

Article 10, Sustainable Use of Components of Biological Diversity, clause (c): Each Contracting Party shall . . . (c) Protect and encourage customary use of biological resources in accordance with traditional cultural practices that are compatible with conservation and sustainable use requirements;

Article 15, Access to Genetic Resources, clauses 1, 4, 5, 6: (1) Recognising the sovereign right of states over their natural resources, the authority to determine access to genetic resources rests with national governments and is subject to national legislation . . . (4) Access, where granted, shall be on mutually agreed terms . . . (5) Access to genetic resources shall be subject to prior informed consent . . . (6) Each Contracting Party shall endeavour to develop and carry out scientific research based on genetic resources provided by other Contracting Parties with the full participation of, and where possible in, such Contracting Parties.

Article 16, Access to and Transfer of Technology, clauses 1 and 2: (1) Each Contracting Party, recognising that technology includes biotechnology . . . undertakes . . . to provide and/or facilitate access for and transfer to other Contracting Parties of technologies that are relevant to the conservation and sustainable use of biological diversity and make use of genetic resources . . . (2) . . . In the case of technology subject to patents and other intellectual property rights, such access and transfer shall be provided on terms which recognise and are consistent with adequate and effective protection of intellectual property rights . . .

Article 17, Exchange of Information, clauses 1 and 2: (1) The Contracting Parties shall facilitate the exchange of information . . . (2) Such exchange of information shall include exchange of results of technical, scientific and socio-economic research, as well as information on . . . indigenous and traditional knowledge as such and in combination with the technologies referred to in Article 16 . . . It shall also include . . . repatriation of information.

Article 18, Technical and Scientific Co-operation, clause 4: The Contracting Parties shall . . . encourage and develop methods of cooperation for the development and use of technologies, including indigenous and traditional technologies . . . [and] shall also promote co-operation in the training of personnel and exchange of experts.

Article 19, Handling of Biotechnology and Distribution of its Benefits, clause 2: Each Contracting Party shall . . . promote and advance priority access on a fair and equitable basis by Contracting Parties . . . to the results and benefits arising from biotechnologies based upon genetic resources provided by those Contracting Parties.

---

try to come to grips with the implications of these two agreements, indigenous peoples and other non-government groups and organisations worldwide are questioning both the ethics and the benefits of their intellectual property requirements.

While the Biodiversity Convention and WTO's TRIPS have the force of international law, governments however have considerable flexibility in interpreting and implementing the intellectual property clauses. Furthermore, the time frame for their implementation is either undefined, or gives leeway to developing country legislators well into the first decade of the next century. And the WTO agreement review is in 1999. It would take a monumental effort, but its intellectual property provisions could be changed. Policy makers, non-government and indigenous peoples' organisations, farmers and environmentalists from all over the world could well influence the direction of national and global policies on intellectual property over living things, and peoples' knowledge about them.

PUBLIC POLICY CONSIDERATIONS—FOUR CASE STUDIES

The following examples illustrate some of the thorny issues that must be addressed by international bodies and policy makers in all countries (like Australia) which have ratified and are now bound by the terms of the Biodiversity Convention and the World Trade Organisation's TRIPS Agreement. They highlight a range of public policy considerations that are being debated around the world, and indicate some of the competing interests that must be taken into account.

Case study 1: Western Australia smokebush patented in the USA

Between the 1960s and 1981, the Western Australia Herbarium assisted a licensed collector for the US National Cancer Institute (NCI) to collect over 1200 plant specimens from WA, for a programme to screen natural products for their cancer-fighting properties. All were sent to the NCI for testing as anti-cancer drugs. Among the specimens were samples of a WA smokebush (*genus Conospermum*). In the late 1980s, the specimens were screened again by the NCI on behalf of the US National Institutes of Health (NIH) (US Patent Office 1997). A compound named conocurvone was extracted and purified from the smokebush and analysed for its effect on HIV. Even in low concentrations, it was found to 'inactivate' the virus (Armstrong and Hooper 1994). Anticipating a significant commercial use, the US—represented by the Department of Health and Human Services—applied for a US patent in January 1993 (granted September 1997, patent number 5672607) involving 24 claims on antiviral naphthoquinone compounds, compositions and uses thereof. Conocurvone was one of these compounds. The following year, the same applicant applied for an Australian patent (granted August 1997, patent number 680872). Like all patents it gave its US owner exclusive monopoly rights to use the compounds from the WA plant, to decide who would be licensed to use them, and at what cost. This case was widely reported in the Australian media, and fuelled a public debate that mirrored growing worldwide concern about who should own, have access to, and benefit from, the world's biological resources.

The patent application prompted hastily-conceived and controversial amendments to the *WA Conservation and Land Management Act*, which now gives the State Environment Minister power to grant exclusive rights to the State's flora and forest species for research purposes (Fourmile 1996). It also led to agreements in December 1993 and March 1994 between the US National Cancer Institute, the Western Australian State government and the Australian Medical and Research and Development Corporation (AMRAD)—a Victoria-based pharmaceutical development company (Armstrong and Hooper 1994; Fourmile 1996). AMRAD was granted an exclusive worldwide licence by the NIH to develop and market a series of anti-HIV drugs derived from the conocurvone (AMRAD 1996), and AMRAD paid the Western Australian government an initial A$1.15 million to ensure access to other smokebush species.

It is not clear what role, if any, the Commonwealth government played in these negotiations. Nor has anyone publicly acknowledged that Aboriginal people have used the smokebush medicinally for centuries (Fourmile 1997). Nobody has revealed whether its prior use by

Aboriginal people led the NCI to the smokebush in the first place; one might well ask. One might also ask whether Aboriginal and Torres Strait Islands peoples will have any protection in future cases where their medicinal plants and knowledge lead to drug discoveries, or whether State governments should have the right to sign away the biological heritage of the Australian continent for private profit.

Case study 2: AMRAD: a made-in-Australia bioprospector and broker

AMRAD is an Australian company established in 1986 for 'the discovery, development and commercialisation of pharmaceutical programmes and projects'. Its 'members' include eleven medical research institutes from five Australian States and Territories (see AMRAD 1996). In 1993, AMRAD established a natural products screening program, and has since 'secured access to biota from some of the most diverse environments in the world—from the wet tropics to Antarctica' (AMRAD 1997a). Within Australia, AMRAD has signed collection agreements with government authorities, research institutes, State herbaria, botanical gardens, Aboriginal Lands Councils and Land Trusts.

The company is collaborating, for example, with the Darwin-based Menzies School of Health Research and the Tiwi people of Bathurst and Melville Islands, on research into 'twelve plants still used as bush medicines' (Sweet 1997). Agreements like this one, and another for the pharmaceutical screening of Arnhem Land Plants signed with the (Aboriginal) Northern Land Council (AMRAD 1995), have sent reverberations through Aboriginal Australia which is still struggling to weigh up the implications of such precedent-setting arrangements, and to assess how best to protect indigenous peoples' rights in the face of corporate approaches (Fourmile 1997).

In the early 1990s, AMRAD signed deals with foreign-based transnational companies, including Merck Sharpe and Dohme (USA), Kanegafuchi Chemical Industry Company (Japan) and Sandoz (now Novartis, of Switzerland) (see Hindmarsh 1994, p. 252). In 1995, it also agreed to screen Australian natural products for the Chugai Pharmaceutical company of Japan (*Australasian Biotechnology* 1996) and in May 1997 announced a five-year, A$15 million, agreement with Rhône-Poulenc Rorer (France) to screen Australian plants for use in the treatment of asthma and related illnesses (AMRAD 1997b). AMRAD (1997a) has also signed an agreement with the state government of Sarawak in Malaysia, for access to Sarawak's genetic resources. By April 1997, AMRAD had tested over 750 000 biological extracts for use in treating such conditions and illnesses as HIV, cancer, hepatitis B, neurodegeneration, anaemia and asthma.

Who, one might ask, will reap the benefits from these agreements—indigenous peoples, other Australians, or corporate coffers? Will the customers of any new medicines be able to buy them at reasonable cost, or will they pay inflated monopoly prices? On what terms were the agreements signed? Will indigenous peoples be adequately recognised or rewarded for their contributions? Will their knowledge and intellectual integrity be protected?

Case study 3: Andean grain patented in the USA and Australia

For centuries, the indigenous peoples of the Andes have grown, eaten and adapted quinoa (*Chenopodium quinoa*) to hundreds of micro-environments in their arid mountain homelands. Quinoa—one of the most protein-rich grains in the world—has been a staple of Andean diets for generations. It went almost unnoticed in the rest of the world until recently, when the food industry started to take an interest in it. The Swiss food manufacturing giant Nestlé has been conducting research on quinoa, and has been granted a quinoa processing patent in the USA.[3] Quinoa has begun to appear in 'exotic' breakfast cereals and pastas, one of a number of 'ancient grains' now attracting the attention of health-conscious consumers in the industrialised world.

This might offer a chance for Quechua and Aymara quinoa producers of Bolivia, Chile, Ecuador and Peru to develop a sustainable export into the markets of the industrialised countries to their north—but not if current trends continue. Two scientists from Colorado State University in the USA have been granted US and Australian patents on male sterile plants of a Bolivian quinoa variety named Apelawa, and any quinoa hybrids produced with them.[4] These patents give the inventors the right to prevent anyone else from making, using or selling quinoa hybrids derived from the patented Apelawa cytoplasm, without permission or payment of royalties. Legally the inventors also have the right to prevent hybrid quinoa from entering the US, if it has been created using their patented technology. The same scientists, incidentally, are conducting hybridisation research on Peruvian quinoa varieties, which, according to one of them, has sparked commercial interest (Ward 1997).

Bolivia's National Association of Quinoa Producers has asked the patent holders to drop the Apelawa patents, and has protested about them at home and at the United Nations. Andean governments are contemplating action. If these patents are not dropped, and if new quinoa patents are granted, the intellectual property rights for commercial development of quinoa in the North will likely wind up in the hands of corporations, and the patent holders will reap the benefits of the patent monopolies over this age-old Andean food crop. The economic implications of this precedent are obvious for farmers of all crops the world over.

Case study 4: indigenous knowledge for the taking?

In the beautiful grounds of the Dreamtime Centre in Rockhampton, Queensland, Aboriginal and Torres Strait Islander guides take visitors daily on walking tours of the Centre's expanding collection of native plants, whose leaves, bark, roots, fruit or flowers have been used since time immemorial by Australia's first peoples—for food, fibre, medicinal and other purposes. Each plant is identified by its scientific name, and by its use. Each reflects the ingenuity and skill of the people who for centuries have studied, described and used the many species of trees and shrubs, vines and creepers which are now displayed in the living exhibit of the Centre's garden.

The Dreamtime Centre's botanical collection is part of an important and growing movement to document, study and preserve indigenous peoples' knowledge of nature. At the moment, however, there is nothing but good will and an ethereal promise of recognition and the 'repatriation of information' under the Biodiversity Convention to stop the world's bioprospectors from simply appropriating these resources and the knowledge that goes with them. Since early colonial times in Australia, European missionaries and botanists have compiled records of the medicinal uses that Aboriginal peoples have made of native plants (Fourmile 1996).

In more recent times, a burgeoning bibliography of books and studies has been compiled to record the unique, and often endangered, knowledge of peoples who are losing their lands, their cultures and their historical relationships with biodiversity. Gene hunters need only pay the price of admission to gardens like the Dreamtime Centre's, borrow the books from public libraries, conduct research on the plants so identified, and then patent their results. They need not give so much as a nod to the people whose knowledge led them to the plants in the first place, and shortened their research time immeasurably. Such a scenario is not just alarmist speculation, it is absolutely consistent with present day trends. The life industries, for instance, are already approaching botanical gardens all over the world, seeking access to their plant collections (RAFI 1996b). Deals are already being struck—with the assistance of a loophole in the Biodiversity Convention which says that all biological collections made before the Convention came into force are outside its scope. Corporations can therefore side-step the spirit of the Convention, which obliges them to negotiate with the source countries of biodiversity, by going instead to cash-strapped botanical gardens whose officials may be more willing to make a deal.

BIOPROSPECTING OR BIOPIRACY?

These four case studies are among thousands of examples from around the world which illustrate what has come to be known as 'bioprospecting'—the search for commercially useful biological resources. Private companies, academics and public sector researchers (increasingly in partnership with a corporate backer) are combing the fields, forests, waters and traditional medicine chests of the world, for useful biological resources. In reality, 'biopiracy' more accurately describes what they have done to date. Resources and knowledge have simply been appropriated from countries and peoples whose permission was seldom sought, and who received little or no credit or compensation for their contribution to bioindustrial developments. A move is now afoot to correct this practice.

The case studies presented above also illustrate some of the pressing public policy concerns that all interested parties must address, including:

- how to assess the impact of national sovereignty over biological diversity on different interests—including various levels of government, indigenous peoples, industry, and public sector institutions;
- how to regulate access to biological diversity, and determine the conditions under which access is granted; defining what is meant by an 'equitable sharing of benefits' and finding ways to ensure that it happens, both nationally and internationally;
- assessing the implications for different interests of intellectual property monopolies over living organisms;
- conserving the knowledge of indigenous peoples' and other communities; protecting the intellectual integrity of indigenous peoples in the face of intellectual property regimes.[5]

CONCLUSION

Australia prides itself on being the 'clever country', and biotechnologists are seen to be among the cleverest of all, leading Australian industry into the new millennium. But just how clever are they? And clever at what cost? The answers will lie in how Australia tackles the policy issues above. By consistently arguing the case for industrial interests, will Australia develop a viable biotechnology industry of its own, or set itself up as a perpetual branch plant to Northern corporations—providing raw materials but rarely owning and controlling production? In aspiring to the big league, will Australia simply duck the ethical questions about whether living organisms (including human genetic material) should be patentable, or will it limit the scope of patent law? Will Australia subvert the collective rights of Aboriginal and Torres Strait peoples under its present intellectual property regime, or use its obligations under the Biodiversity Convention to contribute to a process of reconciliation? In the next very few years we are certain to discover the answers, especially with heightened public awareness and debate about the issues of bioprospecting and biopiracy.

NOTES

1 RAFI (see website—http://www.rafi.ca) has monitored negotiations of the United Nations Food and Agriculture Organisation and the Convention on Biological Diversity during 1995–97. Australia has characteristically aligned itself with the USA, and consistently adopted a pro-industry position.

2 See Internet source www.ims-international.com/press/corp96.htm

3 On 3 November 1993 Nestec (Switzerland) was granted US patent 5264231 for a quinoa processing procedure, and have filed for patents in Europe on the same process. Nestlé has expressed interest in quinoa research being conducted in Ecuador.

4 Duane Johnson and Sarah Ward hold US patent no. 5304718, and Australian patent no. AU 9222922.

5 Marrie, A. pers. comm., 10 September 1997. Indigenous organisations have proposed that Australia develop *sui generis* legislation, consistent with indigenous expectations, to protect indigenous heritage, including indigenous peoples' knowledge of biodiversity. They endorse the concept of 'heritage' elucidated by Erica-Irene Daes, Special Rapporteur of the Sub-Commission on the Prevention of Discrimination and Protection of Minorities, in *Principles and Guidelines for the Protection of the Heritage of Indigenous Peoples* (1996).

5

Disrupting evolution: biotechnology's real result

PETER R. WILLS

Any form of engineering is an attempt to manipulate the world for some desired end. Genetic engineering is no exception. It manipulates DNA molecules which carry hereditary information needed to determine the biology of individual organisms. The concept of genetic information as the real cause of biological phenomena—ranging from enzyme structures, through disease processes to the character of ecosystems—has grown up as molecular biology has advanced during the last sixty years or so. This belief in genetic determinism is now fostered quite uncritically in the public perception.

In undertaking a critique of genetic engineering and its likely consequences, we must look into the scientific assumptions, conceptions, and interpretations of molecular biology. This chapter begins with an overview of some of the basic ideas of molecular biology and then proceeds with a discussion of the relationship between genes and organisms. With this background, we attain a position to make a sensible assessment of the long-term consequences of the genetic engineering currently being undertaken.

All scientific enterprise is based on implicit assumptions which in effect define the 'paradigm' of thinking that the practitioners of particular disciplines use. The paradigm of molecular biology and genetic engineering is termed 'reductionism'—a belief that all phenomena find their ultimate explanation in terms of elementary physico-chemical processes and events occurring at the level of atoms and molecules. Starting around the late 1930s, the advancing techniques in physics and chemistry started to be applied to the detailed microscopic analysis of intracellular structures and processes (Judson 1979). Molecular biologists now wish us to believe that our understanding of the molecular processes that occur within cells, and what this knowledge tells us about organisms, has been determined without any recourse to ideas beyond those which have been proven to be 'true' in physics and chemistry.

Although the reduction of biology to physics and chemistry is something which can in no sense be proven, it is so widely assumed in scientific circles to be an obvious fact that it even forms the foundation of the public regulatory procedures used to evaluate genetic engineering projects (see chapter 3 for a political analysis of this point). The dominant reductionist paradigm of molecular biology has been challenged by the work of theoretical biologists like Kauffman (1993; 1996) and Webster and Goodwin (1996) who have investigated the relationship between microscopic physical processes and the phenomena we observe, especially biological phenomena, in our everyday 'macroscopic' experience. Even when microscopic entities follow very simple rules, interactions among populations of them can give rise to enormously complex 'emergent' macroscopic phenomena, as demonstrated by the 'game of life' (Wuensche and Lesser 1992) or the sandpile model of 'self-organised criticality' (Bak 1996).

Emergent properties are so characteristic of biology and so pervasive in the living world that the task of actually finding explanations based only on explicit molecular descriptions looks impossible—in principle as well as in practice. Seemingly extraneous environmental factors can serve as 'boundary conditions' that have enormous influence on the overall mode of behaviour of complex biological systems, to the extent that the internal molecular workings often do not appear particularly relevant to what is observed. The reductionist viewpoint then becomes more like an article of faith: the laws of Nature and order in the universe are those of physics and chemistry alone; the ultimate reality is comprised of the symmetries and properties of the unified field describing the Big Bang and the subsequent quantum mechanical unfolding of the material universe, including all of biological evolution (cf. Dawkins 1986 and Lewin 1994). By looking at the character of this assumption we will be able to unravel the issue of how genetic engineering is likely to affect the ecological and evolutionary processes that define life in the global biosphere.

MOLECULAR BIOLOGY AND REDUCTIONISM

Knowledge about the molecular mechanisms operating inside cells has exploded since the explanation of the basic modes whereby genetic information is transferred from one material form to another. The simplest, most evident path of genetic information transfer is summed up in what Francis Crick (1958) dubbed 'The Central Dogma' of molecular biology: 'DNA makes RNA makes protein'. DNA is identified as the genetic material which carries the blueprint of the cell. RNA constitutes the copies which are made from the cell's DNA library and transported from the nucleus into the cytoplasm and elsewhere for use in maintaining cellular functions. The proteins inside cells are catalysts—which carry out the vital metabolic functions that keep a cell going—constructed from the information in the RNA. The Central Dogma contains another 'truism' for the molecular biologist: 'Once information has got into protein it cannot get out again' (see Crick 1970). That is, information cannot be fed back along the pathway—protein to RNA to DNA. Interpreting biology in these terms is an easy proposition if one adopts a reductionist attitude. That attitude can be characterised as reductionist on two counts. First, there is the assumption of *material* reductionism:

> Now that we know the physical structure of the genetic material, we may state the aim of molecular biology as defining the complexity of living organisms in terms of the properties of their constituent molecules (Lewin 1994, p. 3).

It is assumed that what we observe comprises a set of material constituents, atoms, molecules and so on, and that change occurs via interactions among these elementary constituents. Although many individual biologists hold personal beliefs in something 'beyond' what can be discovered by using the methods of physics and chemistry, material reductionism is carefully maintained as the framework for intellectual discourse in molecular biology.

The second assumption is that of *genetic* reductionism—that all of the significant physico-chemical features that distinguish organisms are explainable in terms of their genetic constitution. The strongest proponent of this idea is Richard Dawkins, for whom the whole of biology is a mathematical 'space' of organisms represented as nothing but genetic sequences:

> Sitting somewhere in this huge mathematical space are humans and hyenas, amoebas and aardvarks, flatworms and squids, dodos and dinosaurs. In theory, if we were skilled enough at genetic engineering, we could move from any point in animal space to any other point. From any starting point we could move through the maze in such a way as to recreate the dodo, the tyrannosaur and tribolites (Dawkins 1986, p. 73).

All biological phenomena—ranging from the short timescale internal workings of individual cells all the way up to the broad sweep of evolution over the billions of years since the appearance of the genetic code—are thus nothing but a complex game among replicating DNA sequences that compete for control of the resources needed to reproduce. Material and genetic reductionism are assumed to be 'proven' because they seemingly explain why Darwin's principle of natural selection is correct and why its competing theory—the Lamarckian conception of evolution through the inheritance of acquired characteristics—is incorrect. Whereas Lamarck argued that evolutionary progress occurred because organisms somehow passed onto their progeny biological improvements they experienced in their lifetime, Darwin proposed that evolution was simply a matter of the 'survival of the fittest'—the characteristics of organisms are determined by the genetic material they inherit and changes occur only as a matter of the random mutation and recombination of parental genes (see Lewin 1994 for a fuller explanation).

Yet, molecular biology differs from physics and chemistry, upon which it is supposedly based. Whereas much of physics is concerned with abstract theoretical problems and the interpretation of basic concepts—like questions of the type: 'Is quantum mechanical indeterminacy a true feature of the world or just a reflection of our incomplete knowledge?'—molecular biologists act as if crystal clear answers have already been given to questions concerning the kind of reality they are dealing with. All biological explanations, however, are couched in language which refers, either explicitly or implicitly, to functions which can only be defined contextually, usually in relation to whole organisms which cannot be specified completely in terms of their molecular parts. For example, the protein which functions as the oestrogen receptor on the surface of a cell, is only called by that name because other cells produce oestrogen molecules which migrate and bind to those particular proteins on the cell surface. There is nothing intrinsic to any protein molecules which define them as 'oestrogen receptors'. It may be assumed that descriptions of biological phenomena are reducible to descriptions in terms of the most basic concepts of physics and chemistry, but this cannot be proven and the topic is seldom discussed by biologists (Nurse 1997). While molecular biologists have been developing techniques of gene splicing, theoretical biologists have been busy re-analysing the concepts on which biology is based and have—by-and-large—conceptualised biology as something more complicated than simple reductionism would indicate. Most of this theoretical discussion is lost on those who practise molecular biology or genetic engineering because they are seemingly not interested in the interpretation of what they do; they are more concerned with the manipulation of Nature than with its understanding. It is our task to make the relevance of these discussions clear within the context of evaluating genetic engineering, so we will begin with the concept of a 'gene'.

THE CHARACTER OF GENES

If we start with the molecular biological description of a gene, that is, a DNA sequence which codes for the amino acid sequence of a protein, we discover very quickly that a 'gene' is a much more complex entity than is often apparent. A description of a particular human gene is given in Figure 5.1, but this coding sequence is only part of the DNA which gets transcribed into the messenger RNA to be translated by the ribosomes in making the protein product. The whole gene actually consists of three parts found at different locations in the cells' DNA, and the RNA copies of these three parts have to be spliced together to give the 'mature message' which the ribosomes use. The ribosomes need more than just the coding sequence. At a bare minimum the mature RNA must have an extra sequence before the coding region that tells the ribosome where to start translating the message into a sequence of amino acids to make the protein. There are other parts of the RNA which are left over from the splicing process. Other non-coding parts of the message likely function as signals recognised by other proteins which exert control over ribosomal function. Thus, even as a DNA sequence, a full gene is much more than the coded specification of a protein sequence.

We are used to thinking of genes in relation to inheritance and that a gene in some way 'stands for' some feature of an organism so that genetic variation is related to variation in organismic characteristics within a population. For example, within the human population, there is a reasonably homogeneous distribution of individuals who carry a PRNP gene which encodes the amino acid valine (about one-third of the population) instead of methionine (about two-thirds of the population) at position 129 (see Figure 5.1). Based on studies of mice which lack prion protein and which seem to have disturbed diurnal rhythms, we might speculate that the particular version of the PRNP gene that we carry (or combination of versions—we each have two copies of this gene, one inherited from each parent) exerts some minor influence on our sleep patterns, but nothing obvious has yet been found.

Other variations (mutations) in the PRNP gene are associated with inherited neurological disease and it is found that the detailed pathology of the disease, its timecourse and the types of spongiform lesions seen in the brain on autopsy, are characteristic of each mutation. So far, the idea of genes being carriers of individual biological traits seems to hold up. However there is one example of a mutation in PRNP which demonstrates that the situation is not quite so straightforward. An aspartic acid to asparagine mutation (at position 178) gives rise to Fatal Familial Insomnia in carriers of methionine (at position 129) and familial Creutzfeldt-Jakob disease in carriers of valine. Thus, the type of disease caused by a mutation is modulated by what would otherwise seem to be an entirely neutral feature of another part of the gene, demonstrating that the genetic encoding of any individual characteristic

Figure 5.1 Sequences of the human PRNP gene and prion protein

The upper line of text (lower case letters) specifies the normal human PRNP gene (Kretzschmar *et al.* 1986) in terms of the four nucleotide bases adenine (a), cytosine (c), guanine (g) and thymine (t) that are joined to form the DNA chain. The lower line of text (capital letters) specifies the translation of the gene into a sequence of 240 amino acids, abbreviated to the standard one letter code (Lewin 1994, p. 11), that are joined in sequence to form molecules of the prion protein (PrP). Positions of two variations in the DNA sequence are indicated. In about one-third of the human population a 'gtg' codon specifying 'V' instead of an 'atg' codon specifying 'M' occurs at position 129 in the protein sequence. A mutation from 'gac' and 'D' to 'aac' and 'N' at position 178 is associated with inherited Creutzfeldt-Jakob Disease (in the case of 'V' at position 129) or otherwise Fatal Familial Insomnia (in the case of 'M' at position 129).

```
atggcgaaccttggctgctggatgctggttctctttgtggccacatggagtgacctgggc
 M  A  N  L  G  C  W  M  L  V  L  F  V  A  T  W  S  D  L  G

ctctgcaagaagcgcccgaagcctggaggatggaacactgggggcagccgatacccgggg
 L  C  K  K  R  P  K  P  G  G  W  N  T  G  G  S  R  Y  P  G

cagggcagccctggaggcaaccgctacccacctcagggcggtggtggctgggggcagcct
 Q  G  S  P  G  G  N  R  Y  P  P  Q  G  G  G  G  W  G  Q  P

catggtggtggctgggggcagcctcatggtggtggctgggggcagccccatggtggtggc
 H  G  G  G  W  G  Q  P  H  G  G  G  W  G  Q  P  H  G  G  G

tggggacagcctcatggtggtggctggggtcaaggaggtggcacccacagtcagtggaac
 W  G  Q  P  H  G  G  G  W  G  Q  G  G  G  T  H  S  Q  W  N

aagccgagtaagccaaaaaccaacatgaagcacatggctggtgctgcagcagctggggca
 K  P  S  K  P  K  T  N  M  K  H  M  A  G  A  A  A  A  G  A
                        g
gtggtgggggggcttggcggctacatgctgggaagtgccatgagcaggcccatcatacat
 V  V  G  G  L  G  G  Y  M  L  G  S  A  M  S  R  P  I  I  H
                         V
ttcggcagtgactatgaggaccgttactatcgtgaaaacatgcaccgttaccccaaccaa
 F  G  S  D  Y  E  D  R  Y  Y  R  E  N  M  H  R  Y  P  N  Q
                                                a
gtgtactacaggcccatggatgagtacagcaaccagaacaactttgtgcacgactgcgtc
 V  Y  Y  R  P  M  D  E  Y  S  N  Q  N  N  F  V  H  D  C  V
                                                 N
aatatcacaatcaagcagcacacggtcaccacaaccaccaagggggagaacttcaccgag
 N  I  T  I  K  Q  H  T  V  T  T  T  T  K  G  E  N  F  T  E

accgacgttaagatgatggagcgcgtggttgagcagatgtgtatcacccagtacgagagg
 T  D  V  K  M  M  E  R  V  V  E  Q  M  C  I  T  Q  Y  E  R

gaatctcaggcctattaccagagaggatcgagcatggtcctcttctcctctccacctgtg
 E  S  Q  A  Y  Y  Q  R  G  S  S  M  V  L  F  S  S  P  P  V
```

is inseparably entwined with the coding of others, even within a single gene.

We have now discovered what is perhaps the most important feature of genes and that most overlooked by molecular biologists: the biological meaning of genetic information is context dependent—each part depends on all the others, just as the words in a dictionary, looked at in purely linguistic terms, are all defined in terms of one another. No gene in and of itself is the carrier of a single isolated trait. Any trait, like the occurrences of any word in a dictionary, is in fact a set of relationships between a large number of other traits, all of which coexist in a relationship of interdependence. This does not mean to say that there is no pattern to the effect of similar genetic changes in different organisms. In fact, many genetic diseases in humans—including many cancers—can be mimicked in mice by finding mutations linked to the human disease and introducing virtually identical mutations into the genes of mice. It is actually this cross-species regularity in the way genetic changes are manifest as traits in organisms—from mice to humans—and our partial knowledge of the general mechanisms of genetic expression, which make the whole enterprise of genetic engineering possible, but we would be foolish to think we could understand all the complexities of genetics on this basis.

The processes that occur inside cells are linked together in complicated hierarchies, making the effects of changes in genes non-additive, whether the change is made to a single base, a whole gene, or perhaps some associated DNA control element. And, if a genetic engineer takes a piece of DNA from one organism and puts it into another then the overall effect of the transformation depends on the location of introduction of the new DNA relative to genes already there. The products of many genes control the expression of others, often in complex cascades of 'signalling' processes involving the production of proteins that bind to specific DNA sequences and regulate their expression, like the *tat* protein that promotes the transcription of HIV genes in infected cells. The effect of these regulatory proteins is often modulated by their interactions with other molecules present in the cell. In spite of this complexity, the view persists that the effects of altering genes can be understood as if they were somehow additive.

There is a deeper reason why the linear model of genetic information processing enshrined in the Central Dogma ('DNA makes RNA makes protein'—assumed adequate for the purposes of genetic engineering) provides only an incomplete representation of the relationship between genes and organisms. A sort of chicken-and-egg dilemma exists with respect to the genetic code. Genetic information can only be decoded in a cell which contains ribosomes able to translate genetic information, and ribosomes are themselves largely products of the translation process.

And where do cells obtain their initial sets of ribosomes? They inherit them from parent cells. A cell which inherited no ribosomes could not decode any genetic information and would die. The decoding of genetic information is therefore not a linear process—it is circular—a closed loop which has been sustained unbroken in every lineage of biology, stretching back, in some cases, many trillions of cellular generations. DNA can only be thought of as a blueprint for an organism because the organism's vital processes are there in the first place to interpret the blueprint (Wills 1989). There seems no way to get ribosomes—needed for translation—unless you have ribosomes—the products of translation—there in the first place.

Molecular biology has tried to ignore the fundamental circularity of genetic information processing, but the phenomenon of prions is now drawing attention to it because prions manage to reinterpret genetic information in an alternative closed loop whereby they reproduce. Prions are protein molecules which can exist in a minimum of two forms. In vertebrates, production of one form of the protein is healthy, but sustained production of the other form is associated with incurable neurological disease. In yeast and fungi, production of the alternative form of a prion-like molecule allows the organisms to change characteristics and so adapt to some environmental pressures without undergoing genetic mutation. Part of the 'blueprint' for these cells is evidently encoded in structural features of special proteins which have their own decoding mechanism. Cells carrying different forms of the yeast and fungal proteins breed true with their own distinguishable inherited characteristics, even though the cells carrying different forms of the protein are genetically indistinguishable as far as their DNA goes. This illustrates that there are continuously self-sustaining processes—subject to variation completely independent of genetic information—involved in determining biological characteristics and functions at the molecular level.

BIOLOGICAL STABILITY AND INSTABILITY

In contrast with the implicit reductionism of molecular biology, the last ten or twenty years have seen the development of a new field of science which is generally called 'complexity theory'. It encompasses parts of theoretical biology but is broader, seeking to elucidate very general features of systems which adapt and evolve, whether they be physical, biological, social or even economic (Waldrop 1992; Kauffman 1993; Bak 1996).Whatever meaning is given to the term 'complexity', it generally refers to a property of systems that display 'emergent' properties (also referred to above). The emergent behaviour of the system reflects the overall pattern of interactions among the individual parts. Very simple rules of interaction between pairs of individual parts can produce complex system behaviours.

Emergent, complex behaviour only arises in systems whose dynamics are non-linear. To understand what is meant by this, we have to understand the strict scientific meaning of the term 'linear' when it is applied to a dynamic system. In linear systems, each effect is proportional to the intensity of the cause which produced it. Non-linear systems contain feedback loops and there is no longer a simple proportional relationship between cause and effect. The behaviour of systems containing feedback is not as easy to calculate as that of systems whose dynamics are linear. The dynamics of all biological systems are highly nonlinear. At all levels, from the biochemical interactions of a single organism to the relationships of interdependence within a whole ecosystem, the connectivity between individual biological processes is very complex and involves a host of links for inhibitory and stimulatory feedback. So, not only are emergent properties abundant in biological systems, but also emergent properties interact through nonlinear dynamic relationships to generate a multi-level hierarchy of higher order emergent phenomena and the complexity of biological systems appears to continue increasing (Kauffman 1993).

In much the same way as different proteins, enzymes, etc., within a single cell have functions which are related to one another—all of which hold together in some sort of stable manner and maintain the existence of the entire cell—so too do the species within an ecosystem. Each species depends for its existence on the context which is formed by all the other species. The maintenance of characteristic interactions among species in an ecosystem is vital for its continued existence. The different numbers of organisms of different sorts and the manner in which these populations change with time cannot survive perturbations and fluctuations of all sizes, as is true for the molecular composition of cells. The particular sorts of complex adaptive systems which we find ubiquitously in biology seem to have some general characteristics. Ecosystems, like cells, generally have a certain robustness, resilience and ability to cope with change. Yet, they also display great sensitivity toward some imposed changes. These two seemingly contradictory elements are ones which we see manifest and expressed in virtually all biological phenomena. Changes in ecosystems often appear to obey what are generally called 'scaling laws', by which biologists mean that as the scale of a phenomenon changes there is a regular change in the frequency or size of some accompanying effect. One such scaling law applies to the occurrence of evolutionary extinctions of various sizes. Whatever the cause of extinctions, we find that extinctions involving only a few species are relatively common and those involving more

Figure 5.2 Pattern of evolutionary extinctions
(a) Extinction size is measured as the percentage of families disappearing in any epoch and is plotted for the last 600 million years. The data come from the 'Fossil Record 2' palaeontological data base. (b) The number of extinctions of each size plotted against extinction size.

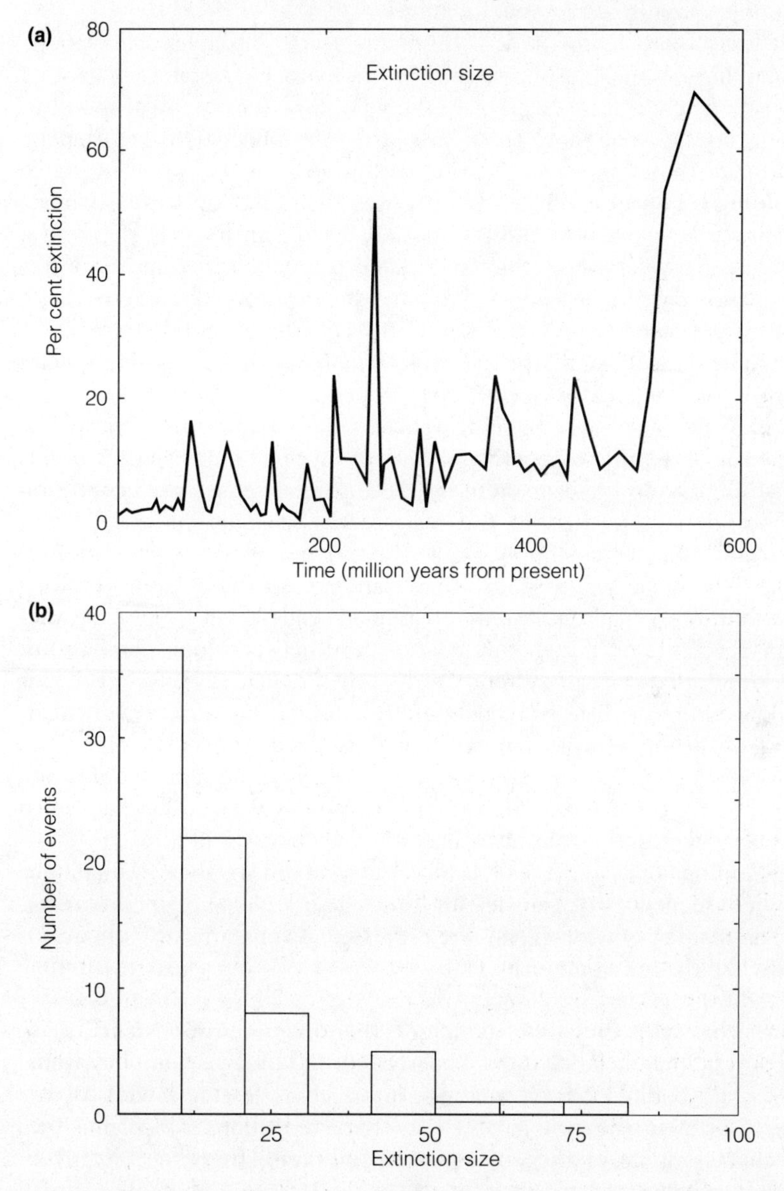

Source: Solé, *et al.* (1997). Reprinted with permission from *Nature*, 388, pp. 764–7, copyright Macmillan Magazines Ltd.

species are rarer. This can be expressed as a regular quantitative relationship called a 'power law' (see Figure 5.2). This scaling law seems to reflect the tradeoff between the robustness of ecosystems and the species which live together in them and the sensitivity that they sometimes have towards small changes.

If we are to evaluate the ultimate effects of large-scale genetic engineering on the biosphere, then we need to have some concept of how genetic change takes place in the wild, how ecosystems respond to genetic change, and how these processes are reflected in the genetic makeup of the organisms occurring within ecosystems. So far, we have not discussed genes much beyond the way they function within a single organism. We must now think about all the organisms that occur in a complete ecosystem, how they are related to one another and how this all depends on genetic factors. We have already seen that a gene is not something which simply gives rise to a trait with all else being able to be assumed equal. Similarly, the evolutionary fitness of any one species depends on what other species are present in the ecosystem and the way in which they are changing. Thus, the idea of 'fitness' cannot be applied in isolation to genetic variants of one particular species unless everything else in the ecosystem is held absolutely constant, a condition which is never satisfied in the real world. Selection pressure is not something that arises simply from 'the environment' to act on single species—or single genes—individually and in isolation. Understanding the evolution of real ecosystems requires consideration of all the interacting species and the dynamic relationships between their populations. We know that an ecosystem is disturbed when a new species is introduced from a different geographical habitat, from another continent perhaps. The entire ecosystem feels the effects. The transfer of organisms across geographical boundaries during processes of human colonisation, has caused gross disruption of a number of fragile ecosystems which had existed essentially undisturbed for long periods of time, in some cases millions of years (Lövel 1997). How should we then evaluate the likely consequences of transferring genes across biological boundaries?

The manner in which genes are expressed as traits in single organisms is very subtle and complicated. If we then extend ourselves to thinking about how the genes in organisms are expressed as characteristics of entire ecosystems and their dynamics, then it is even more difficult to find any simple relationship. The significant 'traits' of an individual species are defined by the manner in which it interacts with all the other species in the system and then the interactions which make up the characteristics of the ecosystem do not arise from the organisms' genes through any straightforward, linear process. Thus, Darwinian selection cannot be viewed as optimising the fitness of any single species except within the changing context of the fitness of the other species, and since 'fitness'—insofar as it can still be defined—reflects the genetic makeup of the organisms that are mutually selected in an ecosystem, individual genes in single species must to some extent encode properties of the entire ecosystem. We are forced to conclude that it is quite impossible to reckon what the effects of some genetic change will be on the ecosystem that the changed organism is found in.

CONSEQUENCES OF GENETIC ENGINEERING

Genetic engineering strategies take no account of the subtle features of biological dynamics that we have just described. Potential side-effects of experimentation and ecological modification may be minimised to the extent demanded by different countries' regulations governing the release of genetically engineered organisms into the environment, but the incremental contribution of any one release to any global process of genetic change due to engineering is treated as if it were inconsequential. An argument can be made for ignoring broader effects if it is assumed that you can take an organism, introduce a new gene into it, and then assess what change has taken place by looking at the effects of that gene on the organism alone. This approach is what is adopted in practice—in countries where there is regulation of genetic engineering. The genes, for example, for human proteins can be transferred to sheep so that ewes produce the human protein in their milk with no other evident effect (Wills 1996).

However, the effects of genes are not always readily evident, as the engineering of genetic 'knock-out' animals—like mice missing the gene encoding PrP (see above)—demonstrates. It is difficult to detect any difference between 'PrP knock-out mice' and normal mice. They appear to live and function perfectly normally, and yet, because there is relatively little difference in the genes encoding PrP in different animals, it is believed that most mutations in PrP must be deleterious and that PrP therefore serves some vital biological function in vertebrates. Looked at the other way around, if a gene encoding a foreign protein is successfully introduced into an organism then certain cells of the organism will then contain that protein. But that knowledge is not an adequate basis for assessing all the consequences of engineering the genetic change. There may be many other subtle effects which remain hidden. When a foreign protein is introduced into a cell it can exert some influence on the way proteins are expressed or function in the cell. The most extreme form of this is the phenomenon of prions which change the manner of their own expression so dramatically that the cell is transformed to a distinguishably different type (Prusiner 1997). In general, the changes are not likely to be so dramatic, they may operate through cascades of interactions with other molecules in the cell and they may only show up when the cells are in some environments; but they may nevertheless be important.

Because one of the main aims of genetic engineering projects is to produce species which can be released into the wild, or at least released for use in agriculture or other commercial activities, we need to assess comprehensively the potential ecological and evolutionary consequences of changing the genetic structure of organisms. Since the advent of Darwin's theory of natural selection, evolutionary change has been interpreted mostly in terms of adaption, the idea being that variant organisms—better adapted to their environment—arise gradually as a result of relatively slow mutation at different points in organisms' genomes. However, as the appearance of microorganisms resistant to antibiotics has demonstrated, important evolutionary changes often involve the introduction of new DNA into an organism from an outside source. It is now well established that there are many paths whereby DNA can cross from one organism to another and bring with it quite dramatic biological effects (Kidwell and Lisch 1997; Shapiro 1997). Also, because the introduction of new DNA into an organism can carry with it the possibility of major biochemical modification—as has happened in the case of antibiotic-resistant microorganisms—it is now thought that DNA transfer is a mechanism involved in the most significant evolutionary changes.

The mode and rate of gene transfer between organisms is probably one of the primary factors controlling the form of evolutionary scaling relationships. Genetic engineering has the capability of disrupting this relationship due to two effects. First, the transfers of genetic material now being accomplished are not occurring within the ecological contexts which have determined their modes of occurrence throughout the aeons of evolution. If there is any functional pattern whatsoever to the gene transfers which ecology and evolution have allowed until now, then humans are about to perturb it in ways much more dramatic than by the transfer of non-transgenic organisms between different geographically isolated habitats. The second effect of genetic engineering is that it could speed up enormously the rate of genetic change in the biosphere, collapsing the timescale defining the frequency at which changes of various magnitudes occur.

Current procedures for the legal regulation of genetic engineering, especially the release into the environment of engineered organisms, are based on assessments of changes induced in single organisms and limited investigation of the interaction between the modified organism and selected members of its proposed ecological niche. It is assumed that any unintended effects can be observed and that any unobserved effects will prove to be innocuous. The wider context is not considered, or rather it is treated as if natural genetic change were a completely random hotchpotch and its rate of no consequence (Wills 1996). The processes of adaption, speciation and evolutionary innovation are thought to lack any regularity, coherence or functional integrity which needs to be taken into account. That is the myth of reductionist biology. In fact, genetic engineering is a technology whereby humans are able completely to change the scale of ecological and evolutionary dynamics and destabilise the mechanism on which the pattern of structures and processes within the biosphere is based.

Genetic engineering has been seized upon as providing a convenient tool for the manipulation of Nature in the service of short-term human endeavours, mostly of an overtly commercial character, whether medical, agricultural or environmental (see discussion in chapter 1). The goal of most of these endeavours is defined by the availability of the means of accomplishment rather than the problem which is supposed to be solved, as has been shown in at least one case study involving the simultaneous production of proteins from transgenic animals and the generation of a pharmaceutical market for their sale (Wills 1996). In this sense it is pointless to look for 'alternatives' to genetic engineering because the technology is by-and-large irrelevant to solving problems that have not already been defined as being amenable to a genetic 'fix'. If the solution to human starvation is perceived to lie in plants whose genes confer the capability of greater grain production, then the alternative to genetic engineering—selective breeding—is likely to be slow, cumbersome and possibly ineffective; but if human starvation can only be overcome long-term by adopting sustainable agricultural practices involving a changing diversity of plant types adapted to local conditions, then genetic engineering will be unlikely to have much of a role to play (as Salleh implies in chapter 12). Genetic engineering can only be used to ameliorate problematic aspects of real world situations as they are perceived by humans and biotechnological 'solutions' must be expected to generate new 'problems' for further application of the technology.

Humanity is faced with a choice which has more to do with collective consciousness about Nature than it does with any question concerning the merits of genetic engineering (Wills 1994). Either we apply ourselves to better understanding and adapting to the dynamic processes of global ecology and evolution or we allow the initiation of a multitude of local perturbations in the constitution of those processes which then propagate and interact, demanding an ever-increasing level of intervention in the service of changing human aspirations. With each new intervention, we will not only make a new selection of human values, but we will clumsily commit future generations to deal with the disastrous consequences of that choice which, in a sense, we have by then encoded in the genes of organisms alive in the environment.

CONCLUSION

The processes for interpreting genetic information are so intricately complex and so diffusely distributed throughout the global ecosystem that it is completely impossible to determine in advance what the ultimate consequences of some even apparently minor genetic change are going to be. If we allow genetic engineering projects to modify the biosphere at a rapid rate and on a large scale, then we can expect to perturb the dynamics of genetic change to such an extent that the changing ecological patterns which have emerged, survived, declined and re-emerged up until our current evolutionary epoch will be completely disrupted.

Genetic engineering will not just be making some minor modification or contribution to the last three billion years of evolutionary development, but rather it will be taking the whole process and putting it into human hands—as if humans can understand it and calculate its effects so well that they can maintain whatever is needed for its well-being and functional integrity. Evolution has been judged thoughtlessly by scientists and governments to be in need of assistance so that its outcome will better suit the aspirations of twenty-first century humanity, and the tools of genetic engineering have been seized upon as the new means of improving or reconstructing Nature.

To arrest this thoughtless pursuit, we urgently need a moratorium on the release of genetically engineered organisms into the environment. All releases of genetically modified organisms into the environment should cease immediately and the ban should remain in place *indefinitely* until all the questions raised concerning the connection between ecological patterns of stability and change—and by implication, evolutionary alteration—have been given scientifically satisfactory answers, in short, have been 'understood'. We are a long way from that and cannot as yet predict when we may arrive at such an understanding.

Part 2

Bioethics, eugenics and risk

6
The quality-control of human life: masculine science and genetic engineering

ROBYN ROWLAND

Subtly, step by step, we are changing the nature of being human and eroding the control women have had over procreation. The latest male-controlled technological intervention—genetic engineering—has the potential to extend the control of masculine science over women's fertility and procreative potential in ways unimaginable a few decades ago. Camouflaging the intention to map and control human genetics with the rhetoric of 'helping the infertile', women are the experimental raw material in the masculine desire to control the creation of life—patriarchy's *living laboratories*.

The product of genetic engineering is 'man-made' [sic] and therefore better than nature; and because our society does not accept the imperfect, women will be placed under more and more pressure to use all technological means offered to secure perfection. Less and less assistance will go to those who make the 'mistake' of having an imperfect child. So in the age of the perfect bioengineered and homogenised product, difference—named 'defect' or 'abnormality'—will become increasingly less acceptable.

The drive for commercial and/or scientific 'success' blinds researchers and developers to the risks and moral implications of their work. Men in power, the makers of ideas and systems of control such as science, construct a make-believe world in which 'free choice' exists, in which individuals supposedly make choices about their lives unhindered by social responsibility to others. Here, the way power works is subtly hidden behind claims for personal autonomy, such as that over procreation. Belief in human control of choice is used in order to reduce human fear of risk. But risk—risk of being hurt, of death, of a handicapped child, of a sudden disability or illness—is in the nature of life. The 'control myth'—the myth that we have choice—leads people to believe that they are free, that there is no need to challenge those in power.

Mary Shelley, who wrote the book *Frankenstein* (1816), saw the perils of the control myth exercised through *biotechnic* intervention into natural processes. Frankenstein, the creator of a monster, lacks the imagination to envisage the implications of this desire to control the creation of life. Thinking only of his own expected glory, he muses to himself: 'a new species . . . a new species would list me as its creator and source; many happy and excellent natures would owe their being to me. No father could claim the gratitude of his child so completely as I should deserve theirs'. Frankenstein's monstrous creation—who seeks only companionship and love—however, eventually destroys those loved by Frankenstein, and in the end the creator himself. In his suffering the monster cries: 'Oh earth! How often did I imprecate curses on the cause of my being! The mildness of my nature had fled, and all within me was turned to gall and bitterness'. In the popular imagination, the monster has taken on the name of his creator, Frankenstein, conflating the scientist with the monstrous. How visionary has Shelley's work been?

HUMAN GENETIC ENGINEERING

To control the outcome of the child-bearing process, masculine science is now developing procedures to create 'perfect', 'unproblematic' people, through sex determination, or through the elimination of genetic illness, or through gene-enhancement of healthy normal people. The dream of quality controlling human life raises profound questions essential to the value structure of our society. It raises ethical, social and legal issues about eugenics, social control, and about the use of women's bodies as raw materials and as experimental laboratories. Any intervention in the structure of the embryo will ultimately affect women, because it is women who will be expected to carry genetically-manipulated embryos through to term, to deliver the resulting children, and to rear them.

Genetic engineering can take place in the form of somatic gene therapy, when new genes are inserted into existing 'defective' cells to replace supposedly faulty genes. The resulting changes are not however passed on to any offspring. Germ line therapy involves changing the germ cells (eggs or sperm) or a fertilised egg, which means that changes to the genetic blueprint will be replicated in the next generation. Micro-genetic engineering is no longer a novelty. It is supposedly being developed for the purposes of 'curing' and 'preventing' genetic diseases and 'enhancing' an already normally developed person. DNA 'finger-printing' techniques are already used in the area of crime detection and immigration. Work has now begun on the complete mapping of the human genome—the biologists' so-called 'Holy Grail'—and it is now almost complete (see Rothblatt 1997).

Molecular biologists and geneticists are developing gene probes to determine the genetic bases of some diseases, such as sickle-cell anemia and thalassaemia, Duchenne's muscular dystrophy, Huntington's disease, cystic fibrosis, Lesch Nyhan disease, Down's syndrome, manic-depression and cancer. In 1990, the first trials for human gene therapy approaches were approved (Joyce 1990). Seven years later, more than 200 clinical trials were underway worldwide—with hundreds of patients enrolled. Although gene therapies are predicted to become in the not-too-far distant future as routine as heart transplants, so far there has been no successful outcome (Verma and Somia 1997, p. 239). Apart from their claims that they will find genes which will cure diseases, scientists believe that they will be able to prevent certain health problems from appearing. For example, they are discussing an early warning system for preventing diabetes, as well as fighting viral diseases and stopping neurodegenerative diseases.

It is disturbing that the research is moving from a discussion of curing genetic diseases to preventing all manner of problems which might arise in children such as asthma or stress. Heart disease and cancer—which have known environmental factors involved—are also considered as genetic problems. Increasingly, health and behavioural problems are linked to a genetic cause. For example, predisposition to problems in the menopause is now discussed as genetically predetermined. This attitude may lead to invasive medical procedures for problems which may not even be biological, let alone genetic. Commercial interests are developing 'kits' for the genetic screening of a person's genetic code for possibly defective genes. It is clear that when translating genetics into a testing procedure, scientists are reducing problems to a simple analysis, looking for single-gene causation of possible multiple-gene defects and ignoring environmental or psychosomatic factors (Rowland 1992; Schmidt 1997).

Once a supposed genetic problem is detected, the question is, what action needs to be taken to restore the person to health? To date, discussions have concentrated on somatic therapy. But some scientists argue that it should be replaced by germ line therapy, since its effects are passed on, whereas somatic gene therapy will only deal with the problems for the immediate individual. Germ line therapy, some claim, will deal with 'cleaning up' the human gene pool in general and will therefore be more 'cost-effective'. As one commentator put it: 'Gene therapy could be used to alter the genetic blueprint of whole populations' (see O'Neill 1991).

Despite the common message from the biotechnology experts that modern biotechnology is user friendly, the Feminist International Network of Resistance to Reproductive and Genetic Engineering (FINRRAGE) argues that gene-technologies are not 'error-friendly'; that is, if errors are made there is no way that the technology can be reversed. With somatic gene therapy, possibility exists of an increased multiplication of cells which was not desired and the development of cancer. Once inserted, the genes are uncontrollable and may embed themselves in the wrong place. The inserted gene may be inappropriately expressed, or may affect other genes around it. Retroviruses which are used to carry the DNA in gene splicing may also be dangerous (Rowland 1992).

Perhaps a more fundamental problem with the work in genetic engineering is that it ties behaviour back to genetics. For many decades now, feminists, and many social and holistic scientists, have been working to encourage a move away from biological determinism, arguing that masculinity and femininity are not biologically-determined, that dark-skinned races are not less intelligent than white, and that the poor are not poor because they somehow biologically deserve it. Science has shown the public that heart disease and cancer are related to stress and to diet. What happens to these variables in the great genetic push? Stress should not be dealt with by changing the genetic structure of a human being, but by changing the environment which produces stress and changing the ways that people learn to cope with stress levels.

DANGERS OF GENETIC ENGINEERING

Health risks have occurred for those working within genetics. Molecular biologists at the Pasteur Institute in Paris developed various cancers while working on techniques of genetic engineering. Two died. In a further incident, farm workers in Argentina were infected with a genetically manipulated rabies vaccine which was being tested by the Americans without the permission of the Argentinian government (Law Reform Commission of Victoria 1988, p. 15).

The approach of scientists to environmental hazards and pollution is increasingly to determine a method by which those likely to become ill through environmental contamination can be screened out of certain workplaces. What seems to be the ultimate insanity of a science created in capitalism is the argument that we do not need to clean up the environment, instead we make people genetically more able to deal with pollution. Scientists have studied genetic predispositions for sensitivity to drugs, pesticides and heavy metals, as well as studying the sorts of people who may have an increased risk of developing cancer, heart disease and allergies (as Gesche also comments upon in chapter 8).

But who will control the products of this genetic engineering and its application? Scientists themselves will not have ultimate control. With commercialisation, that control lies in the hands of the state and multinational corporations. The very fact of patenting animal, plant and human genetic material indicates that its purpose is to generate profit. In 1997, Melbourne pharmaceutical company AMRAD (see also chapter 4) was reported to have patented a mildly retarded child's DNA sequence after researchers had discovered an unusual genetic kink on a chromosome (Dow 1997; and chapter 14). This 'centromere' could be cloned and used to build artificial chromosomes (gene carriers) to deliver gene therapy 'to potentially alleviate suffering from such inherited disorders as haemophilia and cystic fibrosis'. The patent came under instant criticism from Australian Genethics Network director Bob Phelps who said 'that the rights of an anonymous donor had been extinguished and that drug companies stood to make "hundreds of millions of dollars"'. Phelps added that 'there should be a ban on the patenting of human genes because they are "discovered, not invented"' (see Dow 1997).

As genetic testing becomes more effective to predict potential health hazards for people, many companies are intending to develop vaccines and gene therapies to ameliorate or prevent these conditions. As a geneticist at the University of Utah is quoted as saying,

> There are two levels: first, you test with markers to define predispositions, and having implicated [certain] genes . . . you study how these genes work. Then new means of invention will suggest themselves, and you market the drugs you can take to mitigate the predispositions (see Saltus 1986).

Those interested in using predicability tests include various multinational and insurance companies. A study on genetic screening in 1992 by the London-based Nuffield Foundation argued 'that insurers should not be free to cancel existing [health insurance] contracts in the light of new knowledge, and that liberal societies should acknowledge an obligation towards those for whom commercial insurance is declined' (see *Nature* 1992). Today, some of the biggest US companies are using genetic tests of workers' blood and urine to reveal so-called 'susceptibility to harm'. The screening of workers for genetic imperfections shifts the responsibility for workplace illnesses to the worker rather than to those controlling the work environment (Wheale and McNally 1988, p. 257). In a 1993 *Time* telephone poll of 500 adult Americans, some 87 per cent disagreed with the proposition that it 'should be legal for employers to use genetic tests in deciding whom to hire' (Elmer-Dewitt 1994, p. 24).

Testing can further be discriminatory where medical schools might exclude older students because training is costly if the student is 'likely' to have a heart attack and die young. Similarly, airline pilots are another group that could be singled out. Dr Kenneth Pagan from California indicated that though this kind of screening is possible, the opposite could also be true. 'In factory jobs for which training was cheap, but pensions were costly, such jobs could be relegated to people likely to die young'. As early as the 1970s companies began to deny insurance to people who were carrying sickle-cell anemia and yet were healthy (Schmeck 1986). (Bioethicist Nicholas Tonti-Filippini has personally experienced genetic discrimination in Australia—see chapter 7: note 1.)

There is a possibility with DNA fingerprinting and with the work on mapping the human genome that in future, people may be given 'gene credentials' indicating that the person is 'predicted' to develop depression or stress-related disease, or has an inherited disease. When prominent science reporter Robyn Williams interviewed the Human Genome's director he was told,

> . . . 'yeah, it is a possibility . . . they will know an awful lot about people which certainly will have the propensity to invade people's privacy. My feeling about it is that the net outcome will be much more good than harm, but there is a risk of harm . . . You have to make your own choice'. (see Williams 1989)

Predictive genetics could also be used during divorce proceedings to ensure that child custody is not given to the spouse predicted to develop disease. Life insurance companies could refuse to allow insurance for those who indicated such predispositions, or could have a cutoff point at which a person was expected to die. All manner of abuse lies open if this technology becomes more widely available and accepted. Much of this work will be done under the guise of 'genetic counselling' set up altruistically to assist people to make decisions about whether or not they should bear children—and increasingly which kind of child they *should* bear. The Law Reform Commission of Victoria (1988, p. 14) summed up:

> The development of genetic counselling and therapy may raise other problems for patients. Pre-natal genetic testing is likely to become more common, perhaps even routine. Media publicity of birth defects and the new technology available for detecting and treating them will increase patients' demands for pre-natal testing. There will be commercial pressure from the biotechnology companies producing testing materials. Physicians may urge that pre-natal tests be undertaken to reduce the chance of negligence suits. New sub-specialities in medicine, such as genetic counselling and clinical and laboratory genetics are developing to serve the new market. Government financing of public health programmes may focus on prenatal genetic health issues. Medical insurance benefits may be conditional upon pre-natal screening.

Some writers also express concern about the connections between genetic engineering research and military interests. Graham Pearson, Director of Britain's Chemical Defence Establishment at Porton Down, argues that genetic engineering has blurred distinctions between chemical and biological warfare. The concern is that genetically-altered naturally occurring toxins and viruses may be released during warfare. In the US, biological research has increased substantially in the Department of Defence compared to work done under the National Institute of Health.

Some argue that genetic engineering and DNA fingerprinting will greatly advantage society. McLaren (1987) graphically presented the 'pro' case when she wrote about the genetic problems of some children,

> [Some] may expect to live for two or three years (as in Tay Sachs disease), for ten or twelve years (Lesch Nyhan) or for twenty years or more but with rapidly increasing disability (as in Duchenne's muscular dystrophy). Some conditions cripple either the body or the brain (Down's syndrome, for instance) but do not kill. Others do not appear until middle age (Huntington's disease), so people with an affected parent may live all their lives in dread of developing the same disease.

The prospect of invasive genetic work on people who do have genetic diseases raises the issue of our attitudes to disability in general. Special-needs people are seen either as hopeless beings with no quality of life, or as stoically carrying on against all odds. But these people themselves point out that there are a many strengths, weaknesses and capacities in the so-called 'disabled person'; society often stereotypes the problem.

Alison Davis, born with spina bifida, argues that she is not less of a human being because her legs do not work. She argues that discussing a disability in isolation from the individual who has it encourages people not to understand disability but to abort the problem. Davis (1984) argued that one of the reasons that abortion after pre-natal screening is advocated is because it is more 'cost-effective than caring'.

Marsha Saxton, who also has spina bifida, has advocated the rights of disabled people. She argues that the approximately 40 million such people in the US have been 'silenced'. She challenges the following assumptions: that having a disabled child is wholly undesirable; that the quality of life for people with disabilities is less than that for others; that we have the means to humanly decide whether some are better off never being born (Saxton 1988, p. 218).

She points out that most people have not had contact with special-needs people, because of the latter's institutionalisation; this distance has created false stereotypes and negative images. Her advice to a woman considering pre-natal screening with the intention of aborting an 'abnormal' fetus is to ask herself whether she has sufficient knowledge about the specific disability itself, whether she personally knows any disabled adults or children, and whether she is aware of the distorted picture presented to people of the lives which disabled people lead. The major issue involved, she argues, is the availability of societal resources for the parents, the family and the community to assist the child to develop to its fullest potential. She writes, 'we will most likely never achieve the means to eliminate disability: our compelling and more profound challenge is to eliminate oppression' (Saxton 1988, p. 224).

Beck (1995, p. 95) concurs with the view that gene therapy is not just a simple way of removing congenital diseases. He argues that it is, in fact, an attempt to 'abolish the congenitally diseased' and to reduce people's lives to a single characteristic. Future generations are being determined not by dealing with a human body but with 'what seems to be dead matter that can be arbitrarily and 'painlessly' selected and instrumentalised according to certain features'.

But the varying degrees of disability brought about by different genetic problems, cannot be assessed in any way by pre-natal screening. Discussing two of her children who were born with cystic fibrosis, Jackie Andrews indicated many affected children have happy lives and live into adulthood. Her daughter Melanie died at the age of 10 after a year's serious illness and mild disability throughout her life. Her second daughter Alex at the age of eleven had just won the medal for her class in judo. Given the opportunity, Jackie would not have opted for abortion of either of her daughters. She says that 'in spite of her short life, Melanie enjoyed her 10 years . . . and the unborn child, like Alex, may have something to achieve' (see Hughes 1986). The life of the paralysed Christopher Nolan is another example of the achievements possible when loving care and assistance are available. Nolan has produced two widely acclaimed books: his second (Nolan 1987) won the Whitbread Book of the Year Award in England.

Further rebutting the biotechnological vision of the desirability of gene therapy is that certain genetic illnesses can actually produce positive effects. As O'Neill (1991, p. 43) reported:

> Those who develop sickle-cell anaemia have two copies of the defective in their chromosomes. But for those who inherit only one copy of the sickle-cell gene, it confers a better ability to survive malaria—a disease common in those parts of Africa where sickle-cell disease also is endemic.

In her study of women who have used or refused amniocentesis—diagnosis for chromosomal abnormality in the fetus, Katz Rothman (1986, p. 238) cited women who argued for acceptance of the risks in having a child, rather than screening them. Summing up the dilemma she wrote:

> For some people, this is the answer: we accept our children, we comfort them, we do what we can, but we cannot take their pain and their lives for them. For others, that is not enough. For some, it is better not to have a child than to have a child that will suffer.

But the hideous irony is that no woman's decision can ensure this anyway. The screening technology is not failsafe, and cannot in any case guard against disability which is not caused by genetics. In many countries, the main predictors of disability and disease in newborn and young children are poverty and the extreme youth of the mother.

EUGENICS

The desire for human perfectibility leads inevitably to selective breeding. As Jacques Testart—a leading specialist in the test-tube baby business in France—said over a decade ago, 'with genetic progress, the way is open for eugenics' (see Vines 1986). Ten years into the future—with rapid advances in molecular genetics—Wilkie (1996, pp. 133–4) was able to project a rather disturbing futures eugenics scenario:

> Sharon had started asking The Question: Where did I come from Mummy? As she patted her daughter's blonde hair (catalogue number: HC 205) and looked into her perfect cornflower-blue eyes (catalogue number: EC 317), Tracy decided that, in the year 2095, there was nothing to be squeamish about . . . Trevor and Tracy . . . had gone to the Ideal Baby exhibition . . . combed through months of back issues of *Genes and Babies* magazine . . . window-shopped at the Swedish genetic superstore . . . What more could loving parents do for their child? Although, she reflected, it might be better not to mention . . . that Trevor and Tracy had not been able to afford the most expensive high-intelligence genetic profile and had to settle for a cheaper model (catalogue number: IQ 200). Sharon would therefore always be less intelligent than Charlotte next door, whose grandparents had taken out a second mortgage to help purchase for her the genes guaranteeing ultra-high-intelligence (catalogue number: IQ 300).

Eugenics was a term coined by Francis Galton (cousin to Charles Darwin) in 1883. The eugenics movement was very strong in England and North America as well as in Germany and Sweden. In the US in 1910 a Eugenics Record office was established and had two outcomes: compulsory sterilisation laws and the restriction of immigration. By 1931, 30 states had such laws. Compulsory sterilisation applied to 'so-called sexual perverts, drug fiends, drunkards, epileptics and "other diseased and degenerate persons"' (Hubbard 1988, p. 228). By 1935, 20 000 people in the US had been forcibly sterilised. Sterilisation laws remain on the books in twenty States. With the first institute for racial biology established in Sweden in 1922, sterilisations on the basis of eugenics goals peaked in 1944 with 1437 operations. Recent revelations indicate that Sweden continued its sterilisation programme until 1976, by which time social and economic grounds had replaced eugenics grounds for the operations (Butler 1997).

In Australia, eugenics gained many adherents also before the development of the policies of Nazi Germany. It particularly appealed to those who believed that 'poverty and misfortune were the result of immutable weakness of character'. An endowment for mothers was rejected by many on the grounds that it would encourage people who were not desirable parents to breed. 'What the world wants is the sound teaching of Eugenics; healthy parenthood . . . and not the reckless propagating of a diseased and undesirable race'. Unfortunately some feminists were involved in this philosophy. Millicent Preston Stanley in her speech to the New South Wales Legislative Assembly in 1927 advocated legislation to segregate 'mentally defective persons, for the purpose of effecting an improvement in the race-stock'. Statements by the Mothers Club of Victoria followed this line and Mrs Priestley was quoted in the *Argus* on 10 September 1935 as saying, 'It had been estimated that in one generation mental deficiency would be reduced by half if sterilisation was legalised. In Australia there was power to make the race we wished. It should not only be a white race, but a race of the best whites'. A movement supported by well-known feminist Jessie Street encouraged the compulsory exchange of health certificates between people who were going to marry. The Racial Hygiene Association issued posters indicating that preparation for marriage without physical tests at a pre-marital clinic was a sign of blindness. These tests included blood tests, chest x-rays, tests for diabetes and tests for inherited diseases. Women were reminded 'that motherhood was "not an instinct but a science"' (Daniels and Murnane 1980, p. 129).

In Nazi Germany the philosophy extended from the sterilisation of 'mentally deficient' people to their literal extermination, and then on to the extermination of other so-called 'undesirable' groups, including gypsies, homosexuals, and Jews. Doctors involved in this extermination, which Lifton described as 'killing as a therapeutic imperative', managed to reconcile their work with that of their Hippocratic oath as doctors. Astonishingly, one Nazi doctor said: 'Of course I am a doctor and I want to preserve life. And out of respect for human life, I would remove a gangrenous appendix from a diseased body. The Jew is the gangrenous appendix in the body of mankind'. But Lifton (1987, p. 16) is not just talking about doctors through the Nazi period. He writes:

> One need only look at the role of Soviet psychiatrists in diagnosing dissenters as mentally ill and incarcerating them in mental hospitals; of

> doctors in Chile serving as torturers; of Japanese doctors performing medical experiments and vivisection on prisoners during the Second World War; of white South African doctors falsifying medical reports of blacks tortured or killed in prison; of American physicians and psychologists employed by the Central Intelligence Agency in the recent past for unethical medical and psychological experiments involving drugs and mind manipulation . . . doctors in general, it would seem, can all too readily take part in the efforts of fanatical, demagogic, or surreptitious groups to control matters of thought and feeling and of living and dying.

The emergence of genetic engineering has caused similar attitudes to again become visible. A former health systems analyst in the office of the former American Surgeon-General suggested that the existence of mentally deficient people prevents the solution of social problems. He said: 'If we allow our genetic problems to get out of hand, we as a society run the risk of over-committing ourselves to the care and maintenance of a large population of mentally deficient persons at the expense of other social problems' (see Stanworth 1987, p. 30).

Is there however a need for enforceable legislation? A survey of consultant obstetricians in England showed that 75 per cent of them insisted that women agree to abort any abnormal fetus they might be carrying before an amniocentesis was carried out (Stanworth 1987, p. 31). Against this background, how tempting would it be for parents to also abort otherwise healthy offspring who do not fit into a fashionable idea of the perfect child? As Hubbard has pointed out, 'in this liberal and individualistic society, there may be no need for eugenic legislation. Physicians and scientists need merely provide the techniques that make individual women, and parents, responsible for implementing the society's prejudices, so to speak, by choice' (Hubbard 1988, p. 232).

Whichever way it is organised, through legislation or 'choice', the outcome of eugenicist attitudes means selecting humans of value and non-value. Who sets the criteria? And what of the role of women, who ultimately will have to carry genetically manipulated embryos to term or who are given the 'choice' of pre-diagnosing the health of their children? Wheale and McNally (1988, p. 274) write:

> The gap between the Daedalean power of this revolutionary new science and technology and our inability to foresee its consequences creates a moral duty which can only be executed when the utopian ideal of perfectibility, which is embedded in the scientific endeavour, is superseded by one of greater responsibility and accountability.

To date both science and commerce have failed in being both responsible towards, and accountable to, women. Indeed, a gradual softening-up process to move the public towards accepting genetic engineering continues.

RESISTANCE

For those who want to resist the bio-utopian world that patriarchal science is eagerly leading us to, what forms of action are open to challenge, impede or stop it? Are there avenues by which women (as well as men) can reject the equation of personhood with fertility, especially under the impact of genetic engineering and the masculine dream of quality control. Resistance especially lies in self-empowerment of women at both the individual level and at the social level (see Rowland 1992, pp. 274–85). A key area is raising the voice of women and of the community—people need to become more articulate on the issues involved. Information about the issues needs to become more accessible so that we all understand better what science is attempting.

FINRRAGE, created in 1984, with members currently in 40 countries, has developed four main strategies for creating change in this area: the open exchange of information and the development of activism; the development of research into this field to create feminist theory and practice; the publication of information and research to increase public awareness and debate; and, pressure through publicity and lobbying to influence legislation and public debate.

Although scientists profess that they are seeking public debate, when the community creates or seeks legislation to control their work they actively work against it in order to reassert scientific freedom (as Hindmarsh highlights in chapter 3). Scientists feel they are a law unto themselves. They claim the ability to self-regulate, yet the evidence of their actions often does not support their rhetoric about the benefits of reproductive and genetic engineering. Clearly, where there is so much profit to be made, biomedical researchers cannot be the guardians of ethical values (as also argued in chapters 1 and 7). Unless the community takes control through legislation and active participation in decision-making processes, medical research in association with commercial interests will continue to determine the nature of our society.

Today, there is an opportunity to stop the movement which is occurring towards the dehumanisation and commodification of life. It is possible to stop the processes which are turning women into living laboratories and children into 'perfectible' products for sale or exchange. Many share the desire for a more humane vision of the future. A unification of that purpose, a belief in the possibility for change and a concern for future generations, is a solid base from which to counter the thoughtless pursuit of scientific control, and the greed for profit, which reproductive and genetic engineering represent.

7
Human genome/ human rights?

NICHOLAS TONTI-FILIPPINI

It was in 1996 that Madame Noelle Lenoir—President of the International Bioethics Committee of the United Nations Educational, Scientific and Cultural Organisation (Unesco)—asserted that while the life sciences and gene manipulation would be an emancipating factor for humanity, they also imposed substantial risk, placing new responsibilities on societies to deal with them. Societies, she argued, must guarantee that the development of the life sciences—with regard to its ethical, cultural, social and economic consequences—pays due regard to the primacy of the human being and makes a positive contribution to the common well-being (Lenoir 1996).

To meet that guarantee, Unesco (1996) prepared a draft *Declaration on the Human Genome and Human Rights* (henceforth called the human genome *Declaration*) as part of formulating a policy position for member states to adopt in 1998. Its purpose is to ensure that human genetics fully respects human dignity and human rights, and benefits the whole of humanity. It aims to prevent abuse, and affirm the need for democratic debate, the dissemination of information, and the promotion of bioethics. The human genome *Declaration* considers matters such as:

- ownership of the information contained in the human genome in general and in an individual's genome in particular;
- reductionism of personality to genetic identity;
- respect for human dignity and rights in the face of that reductionism;
- protection of the individual with respect to safeguarding the genetic integrity of the human species;
- patenting of human organisms;
- insurance, superannuation and genetic profiling;
- access to employment and genetic predisposition;
- discrimination on the basis of genetic characteristics;
- eugenics;

- confidentiality of genetic data;
- injury as a result of genetic intervention; and,
- research review by independent multidisciplinary ethics committees.

Australian scientists are contributors to the Human Genome Project and are doing research in the area of human molecular biology. Some scientists, like Professor Peter Rowe (of Sydney's Children's Medical Research Institute which focuses on fundamental genetics research) have concerns about current experimentation and developments. Rowe (1996) argues that to avoid significant social disruption we need to confront issues like genetics testing and its potential to discriminate in areas of employment prospects, life insurance, and superannuation benefit. He also suggests that an overhaul of medical education is needed to address the new biological knowledge, its relation to disease, and the way that patients are managed.

Unfortunately, few steps have been undertaken in Australia to address the social and ethical implications of human gene manipulation, especially the protection of human rights in relation to genetic engineering, genetic diagnosis, and genetic identity. Although fully informed about the meetings preparing the human genome *Declaration* the Australian Government was not represented at them. Indeed, the Australian Health Ethics Committee has only very recently begun to examine the issues. Some guidelines on Genetic Registers and Gene Therapy were issued in 1988 by the National Health and Medical Research Council (NHMRC). But, in this case, the guidelines were drafted by life scientists not by a broad multidisciplinary committee representative of public interests. They do not take into account the many ethical problems identified in the human genome *Declaration* (and now acknowledged by many scientists). Moreover, the guidelines were developed without community consultation and did not seek to promote public debate (this paralleled exactly the development of guidelines in other areas of genetics—see chapter 3). Because the NHMRC subcommittee contained no representatives of organisations of people with genetic disabilities, the process of formulating the guidelines was contrary to Article 12 of another declaration—the *UN Declaration on the Rights of Disabled Persons.*

This paper addresses six central issues raised by the bioethical prescription for the future that the human genome *Declaration* represents. It also looks at the way the Australian biomedical establishment has responded to those issues. Finally, it formulates five recommendations to address the serious flaws that are revealed in the current Australian review process of human genetics experimentation and research.

DEFINITIONAL PROBLEMS OF THE HUMAN GENOME

The human genome *Declaration* considers that the term 'the human genome' includes:

> . . . the various aspects of the genetic substrata of human beings . . . both to the total genetic material of humanity and all the genes of every individual. Furthermore, it refers to the genome in its tangible aspect (DNA and RNA molecules) and its immaterial aspect (genetic information). Lastly, the notion refers both to the genetic programme which is the source of the vital functions of every individual, and to the genes that can be detached from his [sic] body. It thus seeks to refer to the values attached to human identity (Lenoir 1996).

This definition is very convoluted. It needs to be 'unpacked' to realise, more precisely, the issues of gene manipulation. As it stands, there are eight meanings or aspects to the human genome including: the genetic substrata of human beings; the total genetic material of humanity; all the genes of every individual; the genome in its tangible aspect (DNA and RNA molecules); the genome in its immaterial aspect (genetic information); the genetic programme which is the source of the vital functions of every individual; the genes that can be detached from the body of every human individual; the values attached to human identity.

Obviously, this set of multiple definitions makes any discussion of issues with regard to the human genome difficult and confusing. For example, when the human genome *Declaration* also refers to the human genome as the 'common heritage of humanity' (Preamble and Article One) to which meaning does it refer? Does it mean that the genes of an individual member of the human family belong to all humanity? Or that all humanity may make use of the individual's genes only in their immaterial aspect (that is, use of the information they contain), or after they have been detached from the body, or as a genetic programme, or as values attached to the individual's identity? Or does it only apply to the 'total genetic material of humanity'?

The phrase 'the common heritage of humanity' (presumably intended to rule out patents on the human genome or parts of it) was objected to by some at the Paris launch of the human genome *Declaration* on the basis that it is inappropriate to introduce notions of ownership, even general ownership, of human genetic information. This is the inherent danger in the reductionism of the concept of the human genome: it tends to be thought of as something separable from people. In reality each person has a biological structure, part of which is the genetic sequencing that partly determines the way the person is and the way in which she or he develops. One might just as well talk of the common heritage of the human brain, the human arm, the human wink, or the human temperament. This is a moral error because it seems inconsistent with our notion of respect for persons to speak of common ownership of any part of our persons.

Turning to the genetic identity reference in the above human genome definition, an attempt to grapple with the apparent reductionism is made in the human genome *Declaration*, Article 2:

a. The genome of each individual represents his or her specific genetic identity.
b. An individual's personality cannot be reduced to his or her genetic characteristics alone.
c. Everyone has a right to respect of their dignity and of their rights regardless of these characteristics.

Yet, with regard to genetic identity, ongoing philosophical debates on identity raise many more questions and ambiguities than any that the Article 2(a) definition above could possibly solve. The issue of human identity instead depends on what is essential to continuous existence of the personality despite physiological and psychological changes.

Further questions are posed by this reference. For example, is it being posited that a change to genetic structure implies the end of one's own identity and the substitution of another? Would any gene therapy change a person's identity? Or would such (potential) changes be continuous with one's (established) identity? These may seem to be abstract concepts to some and not of great issue, but they are indeed central to a broad range of matters. They open another Pandora's box of questions. For example, if I commit a crime as one genetic identity and then undergo radical gene therapy, am I the same person to be punished for the crime of that person who had another genetic identity which pre-disposed him or her to crime? If an embryo is found to have a profound genetic disease and undergoes gene therapy, is it still the same identity which has inheritance rights of the natural parents who are now perhaps only partially the child's genetic parents?

There is also a danger that a person's genetic information will be considered a kind of genetic blueprint, fixed at conception for life. Rather, the writhing mass of DNA and RNA that forms the genetic structure of each cell in an individual's body is altered depending on what type of cell and of which organ or part of an organ the cell is a part. Moreover, environmental factors, illnesses and other factors affect the genetic structure of cells (Rowe 1997).

THE GENETIC DATA THREAT: ACCESS, CONSENT AND CONFIDENTIALITY

DNA information can be stored. DNA fingerprinting information is already being utilised in crime detection. Issues arise in relation to access to that information, particularly computerised access. For example, should law enforcement agencies, insurance companies, superannuation funds and employers have access to DNA banks and registers obtained and kept for medical purposes? Genetic diseases can now be identified before symptoms appear and genetic tests can predict whether a healthy couple may have a child with a genetic disease (see also chapter 8). A genetic diagnosis has even broader implications than an HIV positive result, for example, because it is transferred by inheritance.

Storage of genetic information thus raises profound questions about eugenics control, about life insurance, superannuation and even employment opportunities for individuals categorised by genetic analysis. Storage of DNA information is more contentious than the Australia card debate. It is thus pertinent to refer to the human genome *Declaration* Article 9: 'The confidentiality of genetic data associated with a named person and stored or processed for the purposes of research or any other purpose, must be protected from third parties'.

In June 1991, the NHMRC (1991) approved the establishment of genetic registers. The information for a genetic register begins to be collected when a family member is referred to a register for diagnosis, genetic counselling, predictive testing or management of the condition. The NHMRC's guidelines for collecting DNA information state that *the pedigree and health status of family members* are to be recorded in so far as they are known by the family member being interviewed. The use of the phrase 'pedigree and health status' in a human context is indicative of the level and eugenic direction of the NHMRC's thinking. The NHMRC envisages also that additional information may be obtained from members of the immediate family, more distant relatives, hospital records and health professionals caring for the family. The register would be formed by systematic gathering and cross-checking of information.

The guidelines note that information can be obtained by personal contact with individuals only after obtaining consent which may be given verbally. Information about family members is to be recorded even though consent has not been obtained from them. This clearly breaches the right to privacy under yet anther UN declaration—the *Universal Declaration of Human Rights* (UN 1994) (Article 12)—as well as under the human genome *Declaration* (Article 9).

Guidelines prescribe also the use of collected DNA information in on-going clinical management, in epidemiological research, and in research into disorders such as the probable age of onset of different features of a disorder. In the clinical setting it is envisaged that genetic registers would facilitate communication with families and within families for the purposes of clinical diagnosis, genetic counselling, carrier detection, pre-natal diagnosis and pre-symptomatic diagnosis.

Yet, wider use for the register is likely. The Council predicts that registers may be kept by bodies such as State Cancer Councils whose services may include: a reminder service for clinicians and at-risk individuals when the date for the next examination is approaching, tracing of at-risk individuals lost to 'follow-up', an information service for family members, and genetic counselling.

The recording of health information about family members without their consent, the establishment of a database, and access by researchers, other clinicians and 'bodies such as State Cancer Councils' are matters of grave concern. Genetic databases are a serious matter for civil liberties. They provide an avenue by which the medical profession can control and exert pressure on individuals to accept *medical management* even when they are asymptomatic and quite well. Individuals who have been identified as having a genetic disease are likely also to be placed under enormous moral pressure to avail themselves of the technology.

Part of the rationale for genetic registers is to facilitate a virtual eugenics programme aimed at eliminating genetic disease by identifying carriers and persuading them to be sterilised so as not to have families; or to allow the birth of children—but only subject to pre-natal testing and the elimination of fetuses with positive tests for genetic abnormalities, or to have IVF or artificial insemination from donors, or possibly in the future gene therapy (see also chapter 6).

If DNA information is not protected from third parties, individuals identified for—and informed of—a disease status will become virtually uninsurable and less able to attain disability or death superannuation benefits. In future a doctor's consultation may gravely harm the financial prospects not only of the individual but also of every other family member related to that individual. Moreover, withholding such information from an insurer would likely increase the risk of benefits being withheld.

Any formulation of NHMRC guidelines is a vital issue for open community participation (especially in the context of the UN Declarations mentioned above) but instead the guidelines as they stand were formulated and laid down by a small, elite group of 'experts'. Moreover, and alarmingly, the guidelines are not subject to veto by parliament. An important but 'besieged' avenue remaining for their revision (as discussed below) is the Australian Health Ethics Committee.

GENE THERAPY: GENETIC DISCRIMINATION

DNA or genetic technology is in its infancy. With the development of gene therapies, there are real possibilities not only for treating genetic disease but also of developing treatments for or preventing some cancers. There is scope for more reliable, more effective, vaccines and new r-DNA drugs. There are also issues of safety.

Because gene therapies will involve introducing healthy genes to replace defective genes problems of safety are raised in relation to the 'expression' of the introduced gene and the 'targeting' of the gene to the diseased organs. The need for caution is greater in the development of genetics compared, for example, to the development of new pharmaceutical agents. Genetic material is living material and a problem created may continue to replicate itself within the host human body and even as an infection that may be transmitted to others. The possibilities of causing worse disorders such as cancer or of releasing a new and dangerous microorganism into the human population have been raised. There are also problems of genetic discrimination.

Article 8 of the human genome *Declaration* states: 'No one may be subjected to discrimination on the basis of genetic characteristics and that aims [sic] or has the effect of injuring the recognition of human dignity or the enjoyment of his or her rights on the grounds of equality'. Currently it is not uncommon for prospective employers, banks and other financial institutions and even philanthropic organisations awarding scholarships to require medical records and to deny access to people who are poor medical risks. The growing possibilities for identifying (through DNA information) a disease state, disease susceptibilities, or the probability that a person will develop a disease sometime in the future, greatly extends the scope for discrimination. Significantly, a person who is denied access may often not become ill or incapacitated and might never be so.

A real danger therefore exists that a focus upon genetic factors will result in some people being classified in a manner which excludes them from employment, from education, from access to credit and other financial services, and even from being eligible to marry and form a family. Genetic classification may become a bar to participation in many of the goods of life and society.

Insurance and superannuation companies have been lobbying consistently (but cautiously) for access to genetic information so that they can discriminate. So far, they have been successful; there is currently no bar to them acquiring access to genetic information in Australia.[1]

EUGENICS

The Preamble to the human genome *Declaration* states that 'the applications of genetic research must . . . be regulated in order to guard against any eugenic practice that runs counter to human dignity and human rights'. Members of Unesco's International Bioethics Committee refer to this as 'constructive ambiguity' (Kutujkian 1996). The issue at stake is the matter of genetic selection in regard to family formation and raises many thorny questions concerning genetic screening, sterilisation for genetic reasons, embryo and pre-natal diagnosis, embryo selection, selective abortion and infanticide, and germ cell gene therapy.

There are those who see these matters as individual decisions and matters of privacy, and those who argue that many private discriminatory decisions ultimately constitute widespread discrimination. This is precisely the point of anti-discrimination legislation: one person's individual personal preference—when part of a broader trend involving many people—creates the injustice of discrimination against a whole class or category of other people. In this case it is discrimination against the somehow genetically inferior or in favour of the somehow genetically superior, whether that discrimination is on the basis of disability, disease, intellectual capacity, appearance, or gender.

In this vein, Article 4 of the *Declaration* states: 'The protection of the individual with respect to the implications of research in biology and genetics is designed to safeguard the integrity of the human species, as a value in its own right, as well as the respect for the dignity, freedom and the rights of each of its members'. The concept of 'safeguarding the integrity of the human species' raised concerns with many of those at the *Declaration*'s Paris launch, because its ambiguity implies a eugenic aim. This is because the concept might provide the basis for undertaking 'ethnic cleansing' or, if not ethnic cleansing, then at least the 'cleansing' of those people with genetic abnormality or genetic disability (a point Rowland particularly addresses in chapter 6).

Safeguarding the integrity of the human species is considered a 'value in its own right'. The value is in each individual person and respect for their integrity. The final part of the sentence puts the value of the integrity of the human species against respect for the dignity, freedom and the rights of each of its members. This separation of some sort of reductionist notion of species integrity separable from members of the human family, and as something to be valued distinctly from the dignity of human individuals, seems to conflict with human rights objectives founded upon the inherent dignity of every member of the human family.

Perhaps Article 13 of the Council of Europe's *Convention on Human Rights and Biomedicine* is less ambiguous: 'An intervention seeking to modify the human genome may only be undertaken for preventive, diagnostic or therapeutic purposes and only if its aim is not to introduce any modification in the genome of any descendants'.

Yet, to make matters worse, there is avoidance of a major issue in both documents. This issue is the trend toward elimination of genetic diseases through: (1) controlled reproduction using artificial reproduction techniques in conjunction with embryo biopsy for selection purposes or the use of donor gametes; (2) the practice of ante-natal diagnosis in conjunction with selective abortion; (3) the identification of carrier status in conjunction with policies influencing decisions to marry and form a family, sterilisation (voluntary or involuntary) of carriers; and (4) infanticide of those with undesired genetic features either by fatal intervention or by neglect of reasonable care including the failure to provide adequate nutrition and hydration.

This trend will ultimately affect the human gene pool with fewer people with genetic abnormalities reproducing. The outcome is that some human genes are headed for extinction. Furthermore, this change to the human gene pool need not be restricted to serious genetic diseases. Increasingly, it is genetic susceptibility to disease rather than disease itself that is being identified. There is also a growing scope for selection for or against genetically determined characteristics other than disease states. Height, intelligence, appearance and behavioural traits are just some characteristics which have a genetic component, and which may become the subject of parental or social preference for enhancement once the genetic determinants become identifiable.

These issues raise a number of profound questions including, what sorts of people there should be (Glover 1984); what constitutes disease; what is normality; and, indeed, should genetic disease variations in human individuals constitute part of the integrity of the human genome that should be protected? In relation to the latter, does this question make sense when asked in this reductionist way—as though the genes were separate from the people who are the bearers or instantiation of those genes? Furthermore, whether the individual member of the human family (including his or her particular genetic structure) should be considered an artefact or an icon (Brungs 1983); and whether respect for the inherent worth and dignity and inalienable rights of an individual member of the human family warrant protection before birth as well as after birth (UN Convention on the Rights of the Child, Article 23, UN 1994)? These are serious issues not properly dealt with at present. Finally, it raises a profound question about the right of men and women to marry and form a family (Universal Declaration on Human Rights, UN 1994). Article 14 of the human genome *Declaration* gives some recognition to this problem: 'States must guarantee the effectiveness of the duty of solidarity towards individuals, families and population groups that are particularly vulnerable to disease or disability linked to anomalies of a genetic character'.

A relevant distinction is between the treating of a person who has a genetic disease, and the elimination of the disease. The latter would be effected by ensuring that 'diseased' persons do not reproduce, or that their affected offspring are eliminated by embryo selection, abortion or infanticide. This approach has now become commonplace in ante-natal care and in reproductive technology. While not actually fitting the UN definition of genocide which is limited to nationality, ethnicity, race or religion, the intent would seem to be to destroy groups of people defined by their genetic characteristics in the same ways that Nazis tried to eliminate Jews and white Australians have tried to destroy the Australian indigenous people as a group (see Gaita 1997).

That this may happen with the consent of parents does not alter the fact that the concentration on early diagnosis is for the purpose of elimination of the diseased rather than the treatment of the disease. Why is medicine so desperately concerned that one may not reproduce as naturally intended? What will be the impact of elimination strategies with the 'gene pool' of people having a genetic aberration becoming more identifiable, as the capacity to achieve genetic profiling develops under the impact of the Human Genome Project?

We have a very long way to go before we can satisfactorily answer all these questions. The answers will ultimately affect the directions that genetic research and development will take, our lives, life decisions and freedoms. This dilemma, in turn, raises the question of whether we are equipped enough, whether we have developed appropriate structures, to tackle these issues? Conversely, are we like Pontius Pilate washing our hands while the individualistic and impersonal gene-technological imperative sweeps us along as a mob, directionless, without a collective moral or social conscience, and regardless of the rights and dignity of those innocents whose genetics make them vulnerably, identifiably, different?

INDEPENDENT MULTIDISCIPLINARY ETHICS COMMITTEES

The human genome *Declaration* addresses the above dilemma in the following ways:

> Article 12
> States shall provide a framework for research with due regard for democratic principles in order to safeguard the dignity and rights of the individual, to protect public health and the environment.
>
> Article 15
> States shall recognise the value of promoting, at various appropriate levels, the establishment of independent, multidisciplinary and pluralist

ethics committees to identify ethical, social and human issues raised by research and interventions affecting the human genome.

With these articles in mind, it is obvious that independent, multidisciplinary, reviews of medical research and practice will—short of legal proceedings—provide the only satisfactory means of ensuring that new developments involving human subjects are made answerable to the community. Apart from cost, time delays and inconvenience, the litigation process is hampered both by a lack of access to information about what is occurring and by the concomitant difficulty in gaining hard evidence of abuses.

A danger of contemporary bioethics review structures worldwide is that they tend to become the tools of a defensive alliance of biomedical interests used for the purposes of validating the dominant ideology of their alliance, such as partisan and utilitarian genetic determinism. While a scientist, clinician or research institution experimenting with human subjects without ethical review is an easy target for social criticism, vilification, and political interference, the same individual or institution carrying out the same activity after having first assembled a body of like-minded people to ethically 'review' the activity can proceed with impunity!

In Australia, the NHMRC stipulates that institutional ethics committees (IECs) be established in major hospitals and research institutions for NHMRC-funded projects. The committees are composed according to the NHMRC guidelines. If the composition of an IEC is determined (as most are) by appointment by the institution itself, and composed mostly of people who belong to or are already associated with the institution then, as a community, we can have little confidence in that form of review. The appointment by the institution of token outsiders such as a lawyer, a minister of religion, and a layperson does little to reassure. They are chosen by the institution and so are likely to be chosen for their compliance with the dominant ideology within the institution. In any case their minority influence is unlikely to counter the influence of the numerically superior, in-house, professional medical and administrative opinion (McNeill 1993, p. 6).

Consequently, it would overwhelmingly seem that the dominant function of institutional ethics committees in Australia is to validate—rather than to question—the prevailing ideology within the research institution. While this is not to imply bad faith or a lack of ethics on the part of researchers, the reality of an appointment mechanism which involves the reviewed appointing the reviewers is unlikely to result in an impartial, disinterested, review committee!

In 1995 there was controversy over the fact that some State Family Planning Associations appointed their own ethics committees and these were sufficient to give approval for clinical trials of abortifacient treatments that were not approved for use in Australia. In fact under the NHMRC guidelines anyone can establish an ethics committee. It became obvious that a medical researcher only needs to hold a dinner party with a social worker, a lawyer, and a minister of religion (who might have sent his apologies) and a couple of their spouses and that would constitute an ethics committee under the national guidelines able to approve clinical trials and any other research on human subjects. That controversy led to a Commonwealth Health Department review of institutional ethics committees. The review recommended that appointment to IECs should be an open selection process rather than secret and in-house, and that IECs should be independent of the institution. But little has changed.

Institutional ethics committees are meant to follow national guidelines prepared by the Australian Health Ethics Committee (AHEC). In recognition of the unsatisfactory nature of allowing research interests to dominate ethics review, the AHEC was restructured by the Commonwealth Parliament in 1992 so that it was representative of the community and appointed by a process which included nominations from a broad range of national bodies. Under the legislation the NHMRC is obliged to publish AHEC's guidelines on medical research. However the AHEC remains part of the NHMRC and it has been rendered ineffective by the NHMRC. Adequate funding and its own administrative staff had been deprived to it. References on ethical issues have been given to other less community representative committees to prepare guidelines. To sum up, the will of the Parliament has been defeated by the bureaucratic control exercised by entrenched medical research interests.

RECOMMENDATIONS

To address the requirements of meeting democratic principles and independent review specified in Articles 12 and 15 of the *Declaration on the Human Genome and Human Rights*, five goals need to be considered in determining how appointments to committees of review are made, from what categories of person they are drawn, and how the review process may be constructed.

1 *Impartiality*
The composition of review committees should be either of members chosen (like a jury) for their impartiality and disinterestedness, or of an overall impartiality achieved by including a range of differing views and interests that provides a balanced representation of the community and those affected by the type of decisions the committee makes.

2 *Representation*
Review committees should have a capacity to be adequately informed of developments and likely developments in health research and practice, and of the variety of concerns that the community and those who may benefit or be at risk have in regard to the developments. In the area of genetic research it is important that those groups of people whose families are affected by a genetic disease be represented. Committees should ensure that the evidence of their expert witnesses or expert members is not unbalanced.

3 *Responsiveness and democratically conducted*
Review committees should ensure that decisions are responsive to community concerns, the interests of research subjects and clinical practice, and to the practical dilemmas of researchers and practitioners, and that elitist, autocratic and alienating decisions are prevented. No interested group or individual should be permitted to dominate the committee process.

4 *Accountability*
Review committees must be accountable for their actions. Responsibility must at the end of the day rest with those whose democratic task it is to formulate public policy—the elected representatives.

5 *Openness*
Review committees' processes must be transparent to the community and the opportunity for informed, open and interactive debate in the broader community facilitated and encouraged.

NOTES

1 For example, I have a serious disease of unknown origin and which might have a genetic basis. Not only have I been refused insurance and superannuation pension entitlements, but my children are not insurable. In regard to financial institutions, even though I work full-time and had more than sufficient income to cover a mortgage, I could only take out a mortgage with a guarantor because I am dependent on life support. I have even had to fight to retain a driver's licence even though my physician certifies that my condition is stable and as a driver my health does not pose a risk. The Road Traffic Authority was not concerned about my capacity now but what it might become and the same logic might be applied once there is more genetic information available.

8
Genetic testing: a threat to privacy

ASTRID GESCHE

> Jackie and Emma are sisters. They recently took part in a research program looking for alterations in the gene *BRCA1*, that confers a susceptibility to breast and ovarian cancer. They learned that they both carried the altered form of the gene, which put them at an increased risk to develop the cancers. Subsequently a mammogram revealed that Emma had a small cancerous lesion in her breast, while no cancer was detected in Jackie. Both sisters decided to undergo a bilateral mastectomy in the hope that this would reduce the risk for breast cancer. Jackie and Emma felt strongly that they benefited from knowing this genetic information. However, they were fearful that their genetic status could be used against them and their family by insurance companies and employers (cited in Hudson *et al.* 1995, p. 391).

The science that studies genes, how they work, how they are passed on from parent to offspring, and what effect they have on an organism, is called genetics. Genetic factors guide most human characteristics, from our outward features and behavioural attributes to our susceptibility to disease. Traits such as eye colour and height, personality and intelligence, health and disease, are all influenced by our genes. Genetic differences can have a variety of implications for individuals and for society. The story of Jackie and Emma above is one example—it is a story of our time and of our future.

The ability to study diseases at the gene level is enabling us to understand their origin and development to an extent not possible a few years ago. Today, we know of some 6000 conditions that are partly, or entirely, the result of abnormal or missing genes, or abnormal chromosomes. For the majority of genetic diseases, there is currently no treatment and no cure. Some people, like Jackie and Emma, take radical measures to reduce the risk (but there is no guarantee that it is reduced). The sisters based their decision on the nature of their mutation.

Sometimes, however, simple life-style changes suffice to modulate the effects of genetic aberrations.

Over the last few years—and due to the multi-billion dollar international 'mapping' project of the Human Genome Project (HGP)—we have found out a great deal about genes. The HGP started in 1990. It set out to identify all the human genes in a genome and to construct a detailed map of their three million DNA base-pairs located on our chromosomes. The project is now almost complete and the emphasis is rapidly shifting from gene structure and gene sequence to gene function. On the way, it has given us astonishing insights into the 'private lives' of genes (Love also addresses this latter aspect in chapter 9, but in a satirical manner).

Even though geneticists have learnt a lot about the molecular origin of a particular disease, they often have no clear idea of the extent to which genes are responsible for disease symptoms or of how to cure genetic abnormalities. Therefore, many geneticists, scientists, physicians, ethicists, and others ask whether people should be tested for particular genetic predispositions at all. Do individuals really benefit from the information? Genetic information can cause anxiety, stress, and feelings of guilt. Could the information cause more harm than good? Could unfavourable genetic information also lead to discrimination and stigmatisation? Jackie and Emma were fearful that information about their mutation, if not kept secret, could have serious consequences for their professional and private lives. The sisters' story points to unsolved ethical dilemmas: how to keep our genetic make-up and predispositions private and away from prying eyes; and how to ensure that others do not misinterpret and misuse the genetic script of nature and cause us and our families untold harm.

This chapter summarises some of the problems associated with privacy and genetic information in the medical context (others are spelt out by Rowland and Tonti-Filippini in this volume). Essentially, there are many good reasons why legislation is needed to protect our genetic privacy. Legislation should protect medical records; address the sharing of genetic information and the consent process; prohibit the collection or analysis of genetic samples without consent; and, guide and protect the collection, storage, access and destruction of genetic information.

The field of genetics (and, especially, molecular genetics) is developing rapidly, with new developments being reported regularly by the media (as White elaborates upon in chapter 2). New techniques are being discovered, and applications made, in a bewildering array of settings. The following is a signpost to a genetics future still open to many possibilities.

THE SPECIAL PLACE OF GENETIC DISEASE IN A PERSON'S MEDICAL HISTORY

Family members share a common gene pool. At times the inheritance of traits can bring joy, but occasionally they bring sorrow. As researchers probe deeper and deeper into the world of genes, they discover that many more diseases than might have been expected appear to have a genetic component. Jackie and Emma's radical actions were one consequence of this expanding knowledge. Their life-threatening predisposition to breast and ovarian cancer could only have been exposed so early with molecular technologies. Modern genetics may have spared them an early death. But why the fear? What is so special about genetic information?

Several unique aspects of genetic disease distinguish it from other types of diseases. First, there is the special mode of transmission and the permanency of a mutation. Genetic diseases are transmitted from parent to offspring. They are not passed on from any unrelated individual to another—as in the case of infectious diseases. Genetic aberrations, if expressed, do not respond to treatment as do (most) contagious diseases. Mutations remain with the affected person for life.

Then there is the *predictive* power with regard to relatives. For families in which at least one person is affected by a genetic disease, the chance of recurrence in other members can often be predicted by statistical probability studies. In the case of Huntington's disease, a person born into a family with a known history of the condition, has a 50 per cent chance of carrying the defect. Huntington's disease is a progressive, debilitating, disease of the central nervous system. It is characterised by involuntary movements, loss of motor (voluntary) control, and dementia.

Another genetic mutation—cystic fibrosis—affects about one in 2500 Australians. Among the classical features of cystic fibrosis are chronic abnormal secretions of a sticky mucus that blocks the lungs, pancreas and other parts of the body, and often causes an infection of the lungs. The genetic defect in cystic fibrosis lies in a gene called *CFTR*. The defect is unnoticeable in a carrier. In carriers only one of the two inherited copies of the gene is defective. The defect is masked by the second healthy copy, and such individuals do not suffer from the condition themselves (and, as Rowland points out in chapter 6, this can be positive in another way). The defective copy can, however, be passed on to their offspring. Thus, if two carriers produce children, each child has a one-in-four chance of being afflicted with the disease.

In the case of many genetic mutations, we cannot determine when and whether the disease will develop. Unlike Huntington's disease or cystic fibrosis, some abnormalities might not manifest themselves until additional elements come into play. In some patients with a predisposition to lung cancer, abstinence from tobacco and other air pollutants might reduce or totally alleviate the risk. In arteriosclerosis, in which more than one gene is implicated in the development of physical symptoms, many non-genetic factors—such as diet, exercise, stress, and smoke and alcohol intake—influence the outcome of the condition. This means that a person carrying a genetic abnormality can only be said to have a propensity to develop a condition, but other factors—such as environmental influences or other gene sequences—can modify, or even eliminate a negative outcome.

Lastly, there is the predictive power of genetic information with regard to a person's health. The descriptions above of Huntington's disease, cystic fibrosis, and breast cancer, illustrate that the predictive power of genetic testing is substantial, and that the test result can concern others apart from the individual tested. It will have a profound effect on the way we view ourselves and how others respond to us and our family. But the power of predictive testing should also concern the (apparently) healthy. The Human Genome Project is indicating that many common illnesses have a genetic component. They are the so-called polygenic diseases that range from cancer to cardiovascular disease, from asthma to psychiatric illnesses. Since genetic alterations also happen to be the most common causes of illness in the Western world, it is understandable why more and more people are urging scientists and policy makers to protect the privacy of genetic information.

GENETIC REGISTERS, DNA DATABANKS AND LEGISLATIVE PROTECTION

In the medical sector, computerisation of medical records and networking of computers is widespread and increasing. We are now seeing the first medical information services on the Internet. Are the records safe? Hackers have already penetrated hospital and research computers; intercepted the transmission of data; changed patient data; accessed confidential computer files that contain sensitive information about patients and their disease; implanted computer viruses and deleted valuable research records (Eder 1994, pp. 38–42). Their actions pose dire—at times life-threatening—consequences for patients. Criminal tampering by hackers constitutes a serious breach of the code of ethics on privacy and confidentiality that exists between patient and medical professional.

Many countries—among them Canada, Germany, France and Great Britain—have data protection agencies that provide safeguards against misuse of information held in government agencies. Australia, too, belongs in this category. Data collected and stored in Australian government agencies are protected by government legislation. However, Australia has no coherent privacy laws to cover computerised data collection and storage in the private and state public sector (O'Connor 1996, pp. 12–13). What makes matters worse is the fact that common law in Australia does not recognise violation of privacy as a tort (that is, as a civil wrong or injustice) (Lloyd 1997, p. 26). This makes the patient very vulnerable and very reliant on the existing, but unenforceable, codes of ethics associated with various groups of health professionals and para-professionals.

The confidentiality of patient information is the foundation of trust in any doctor–patient relationship. Sometimes it is necessary for a doctor to pass on personal patient information to other health care professionals who are involved in treating the patient. If a patient does not consent to the passing on of information, the doctor usually respects the patient's wishes after due counselling and after explaining the likely consequences of such a refusal. This may be acceptable as long as the patient is informed and competent to make a choice knowing full well the possible adverse consequences to himself or herself alone. However, when the health and future of other genetic relatives are also implicated, nondisclosure becomes problematic. The nature of genetic information is such that it comprises information not only about a patient's present and future physical and psychological health, but also points to the genetic status and future health of members of the wider family.

For a doctor, respecting the privacy of information about a patient is voluntary. The Australian Medical Association's (AMA) Code of Ethics—despite its revision in 1995—does not impose a legal obligation on the doctor to adhere to that code. No statutory restrictions exists on the exchange of sensitive medical information from doctor to doctor, although the code does insist that patient consent must be obtained beforehand.

Medical records are not the only place where genetic information can be found—it is also held in genetic registers and in DNA databanks. In the clinical and research context, genetic registers and databanks store health information about families with certain genetic illnesses. They can also store DNA samples. The primary function of these registers is to collect epidemiological data about certain genetic conditions in order to assist in the diagnosis, counselling, and treatment of at-risk individuals. They also collect, register, and store samples which can provide researchers with biological material for their research.

Most medical research projects, including genetic research, are conducted under the umbrella of the National Health and Medical Research Council (NHMRC). In its *Aspects of Privacy in Medical Research* (1995), the NHMRC requires strict adherence to its framework for handling private information. The overall premise is that confidentiality is of utmost importance—that data are not to be released to third parties without consent. If they are to be released, they can only be used for the purpose for which they were collected. The NHMRC further asks the ethics committees of individual institutions to scrutinise and oversee a project from its design stage onwards (but as Tonti-Filippini points out in chapter 7, serious problems are posed by the composition and value-systems of expert committees). The aim of the ethics committees is to try to strike a balance between the privacy rights of the individual and the health interests of the public. The guidelines however cannot be enforced in law. As long as monitoring remains based on the goodwill of all parties, any infringements of the privacy of individuals cannot be fully pursued legally.

If genetic registers and DNA databanks could be protected from abuse, and if they could be managed in such a way that the individual and the family donating the material would remain the absolute custodians of that information, these depositories could well be an invaluable resource for counselling, treatment and research. At present, Australia is still attempting to achieve that goal. A Canadian study in 1995 revealed problems and substantive deficiencies arising from the lack of formal and uniform DNA-banking policies (see Verhoef *et al.* 1996, pp. 5–16). The study found that most DNA banks lacked written internal policies and written agreements on how to deposit genetic information. This, the authors argued, might lead to future misunderstandings with depositors or to unforeseen legal liabilities.

One consequence of the expanding knowledge derived from the Human Genome Project is that the demand for genetic information will increase in the future. Already researchers look to other potential resources that could supplement existing DNA databanks. By far the largest such resource could be the tiny samples of blood taken from newborns, dried and then used in Guthrie tests. These samples constitute a vast repository of genetic information with an enormous re-testing potential (because of the high temporal stability of dehydrated DNA). For example, in Australia, the Royal Brisbane Hospital stores Guthrie cards dating back to 1986. In other laboratories the samples stored are much older. Eleven respondents in an 1993 American survey of hospitals reported that they had been saving their Guthrie cards since the 1960s. Seven laboratories had amassed in excess of 500 000, with four of these reporting collections of between one and five million cards, and one laboratory six million cards (McEwen and Reilly 1994, p. 197). Such storage has generated additional concern among some people with regard to the privacy of medical records.

Considering the growing interest of stored Guthrie cards, written, legally binding, regulations are needed urgently. They need to accord stringent privacy protection and assurances as to the proper management of these registers, including written regulations on third-party access to the retained Guthrie cards. As long as insurance companies vie for access to genetic information, the public has to be on its guard. For example, should Guthrie samples ever be reclassified as 'genetic test' material, the assurance by insurance companies that they want the results of only prior genetic testing for new applicants (rather than the routine screening of all new applicants) would take on a very different meaning! Everyone born from the mid-1960s onwards would fall into this category of prior genetic testing.

GENETIC PRIVACY AND INSURANCE

Life insurance companies everywhere have been quick to realise that genetic testing for disease susceptibility could be to their advantage by providing an additional criterion in their assessment of mortality risk. The medical and life insurance system in the US is largely built around private insurance companies, many operating in close association with employers. Thus, in the US, the potential for discrimination in health and life insurance coverage on the genetic basis of common disorders can have devastating consequences for affected individuals (as also highlighted by Rowland and Tonti-Filippini in this volume). One early—and well-publicised—case of discrimination in America has been that of Theresa Morelli, who applied for disability income protection with a large insurance company. Theresa was a young, healthy law student of the University of Akron in 1989. Her father had been diagnosed earlier with Huntington's disease. Because Theresa had a 50 per cent chance of having inherited her father's condition, she was denied insurance coverage when her underwriter learned of her father's diagnosis. Even though it turned out later that her father's diagnosis was incorrect (Quade 1993, p. 14), her case highlighted the dangers of genetic testing with respect to insurance cover.

In 1996, the US Congress passed the *Health Insurance Portability and Accountability Act.* To protect against discrimination, it stated that genetic information was not to be treated as a 'pre-existing condition in the absence of the diagnosis of the condition related to such information' (cited in Rothenberg 1997, p. 1756). Laws on the use of genetic information for insurance purposes exist in some US States, but not in others. So far, some twenty US States have passed legislation prohibiting the *use* of genetic testing for assessing access to health insurance. The above Act of Congress, however, does not provide any privacy protection because it does not prohibit insurers from gaining *access* to genetic information. Thus, Jackie and Emma—and indeed most of us—have good reason to be concerned.

In Australia, every permanent resident or citizen has a right to what is largely government-funded health insurance. The problem in Australia, therefore, is not access to health insurance, but a lack of a formal code of practice for 'collecting, storing, using, disclosing and disposing of genetic testing information' (O'Connor 1996, pp. 70–71) and a lack of formal regulation of the sharing of information among the various health insurance providers.

As for life and superannuation insurance cover, the picture is more complex. Like any other business, insurance companies set out to maximise their profits and minimise their risks. They work on the principle that people pay their premiums according to the risk they bring to the insurance fund. Up to now, the primary risk classification for life insurance coverage has been age. Yet in each age group there are individuals with a lower or higher risk of becoming ill. The insurance companies would, of course, prefer to cover more low risk than high risk people, even allowing for the fact that high risk people often pay higher premiums.

Insurance companies maintain that they need access to genetic information because it would help them to calculate a price that fairly reflects the level of mortality risk posed by a group of people (Pokorski 1995, pp. 13–14). Genetic tests, they reassure us, would be used in the underwriting process in a similar way to other clinical tests such as cholesterol testing. Already, insurance companies in the UK reserve the right to have access to the results of previous tests (Masood 1996, p. 389). In Australia, insurance companies have put forward a similar proposal. In Germany, too, previous genetic test results may be requested where the policy exceeds a certain amount of money (Masood 1996, p. 389). In general, insurance companies point out that for them diagnostic genetic tests are just another piece of information useful for the underwriting process, that adds to other factors such as age and lifestyle.

In contrast, critics argue that there is a difference between *knowing* one's predisposition to a disease and *disclosing* information about one's medical history. In general, genetic tests are far more accurate predictors of lifespan than a medical history. In addition, since no person is 'responsible' for his or her genome, people should not be penalised with higher premiums for conditions over which they have no control. There is also doubt that underwriters are versed well enough in genetics to interpret and handle genetic information responsibly—particularly where such information pertains to conditions such as cardiovascular disease or breast cancer in which several different genes can be implicated in the development of the disorder. Even if a person has a sophisticated knowledge of genetics no one, at present, can predict with any certainty who will develop what kind of polygenic disease and when.

It is not surprising, then, that there is a general public drive to prohibit insurance companies to access genetic information. Policy makers in many countries are paying close attention to the arguments on both sides. A clear majority of people do not want insurance companies to have access to genetic data. A public opinion poll in the US conducted for the American Council of Life Insurance (ACLI) showed that 77 per cent of those surveyed in 1994 said that genetic data should not be disclosed (Masood 1996, p. 391). But countries have been slow to legislate, and even greater caution has been taken by others. France and the Netherlands have imposed a temporary moratorium on human genetic testing, while Belgium, Austria, and Norway have opted for an indefinite ban. Even in countries that have moved slowly up to now, the tide is turning against the insurance companies, as public knowledge about the dangers increases.

In Australia, where privacy protection exists at the Commonwealth level, but not necessarily at the State or private level, the Privacy Commissioner took the view:

> . . . that protection [of privacy] could best be extended by giving legislative force to general principles flexible enough to be adapted to the full range of private sector contexts rather than by detailed requirements specific to a particular industry or type of information. Application of the principles to different industries or sectors could be achieved through codes of practice developed in consultation with stakeholders . . . (O'Connor 1996, p. 65).

At present the Life Investment and Superannuation Association of Australia (LISA) follows its own policy on how to handle the use of genetic information and genetic test results. The draft policy states that new applicants are not obliged to have genetic tests for insurance purposes, and aims to ensure the applicant that insurers will maintain confidentiality on all genetic data they may receive. In addition, that insurers will only use the data in the assessment process for that particular individual and not for relatives of the applicant should they also apply. However, the insurer would retain the right to exchange data with other insurance companies that may also be involved in assessing the applicant. In return, insurance companies would ask applicants to provide comprehensive medical information so that they can accurately and fairly calculate the risk of insuring the applicant. In fact, the applicant has a legal duty to disclose that information to the insurer.

While these recommendations are a commendable first step, a

number of issues give reason for concern. First, there is the imbalance of a non-binding code of conduct for the insurance industry against a legally binding duty of disclosure on part of the applicant. Second, life insurance companies have no written code of conduct that deals with the appropriate way of how to manage the collection, storage, and handling of genetic information. In the absence of any binding code of conduct, data protection and misuse of genetic information cannot be ruled out. A last area of concern with regard to genetic testing pertains to workplace practices.

GENETIC SCREENING/MONITORING AND THE WORKPLACE

For employers, a healthy workplace equates with greater productivity, less absenteeism, and lower insurance premiums. Thus, any genetic information that enters the workplace has both substantial privacy implications and a potential discriminatory influence on employment opportunities and promotion. Employers who become aware of a genetic disability in a prospective employee may, for entirely rational reasons, be reluctant to hire the applicant for fear of that worker becoming ill prematurely or seeking disability compensation. But some critics argue that this is not the only reason that employers welcome genetic testing. They may increasingly embrace genetic testing because of its potential to shield them from claims for negligence and liability in occupational health and safety cases (Draper 1991). For example, by selecting and hiring workers who are 'resistant' to harmful substances in any particular work environment, employers may avoid taking remedial actions to minimise environmental risks within the workplace (as Rowland also points out in chapter 6).

In the USA and Great Britain, genetic testing has already started to filter down into management and recruitment practices. In 1995, to avoid workplace stigmatisation based on genetic information, the US Equal Employment Opportunity Commission (1995) issued guidelines to protect individuals subjected to discrimination where genetic information relating to illness, disease or other disorders had been used against them. Later, members of the Committee on Genetic Information and the Workplace of the National Action Plan on Breast Cancer and the US National Institutes of Health–Department of Energy Working Group on Ethical, Legal and Social Implications of Human Genome Research published a number of recommendations addressing the emerging issues on genetic privacy and discrimination (cited in Rothenberg *et al.* 1997, pp. 1756–7). In brief, they recommended:

- the prohibition of using genetic information for the hiring or termination of employment;
- the prohibition of requests for or disclosure of genetic information for employment purposes, including conditional employment offers;
- the restricted access to genetic information for employment organisations;
- the prohibition of release of genetic information without prior written consent of the individual—each application for disclosure must inform the individual to whom the disclosure will be made;
- strong enforcement mechanisms, including a private right of action, for any violations of these provisions.

In Australia and New Zealand, the indications are that genetic testing has not yet found its way into the workplace to any appreciable level. However, as genetic testing becomes more well known and cheaper to perform, this may change. As the range of tests becomes wider and individual tests become more reliable, there is no reason to believe that genetic tests will not become an option for Australian and New Zealand employers.

In Australia, in its *National Model Regulations for the Control of Workplace Hazardous Substances*, the National Occupational Health and Safety Commission (1994) has recommended standards for carrying out health surveillances. These regulations refer to genetic monitoring in the workplace. They aim to ensure that the medical records generated should remain confidential and in the hands of the supervising registered medical practitioner. However, since there are currently no uniform regulatory mechanisms applying to personal genetic information in the workplace, the Australian employee is left with little privacy protection and without the freedom to reject genetic testing should the employer request it.

Is it reasonable to ask perfectly healthy people to undergo genetic testing to secure employment, even though they do not exhibit any physical sign of any disorder, and perhaps never will? Would the applicant not feel pressured into consenting to the testing in order to secure employment, especially in a climate of high unemployment? What happens if the person's job application is unsuccessful? To whom do the data belong? How confidential do they remain? As O'Connor (1996, p. 86) stated:

> Since the employer is often in a position of power relative to the employee or job applicant, unregulated use of genetic testing by employers clearly poses threats to the privacy of employees. As testing becomes cheaper and applicable to a wider range of conditions, the incentives for employers either to obtain the results of previous tests or to seek new tests for their employees will grow.

CONCLUSION

Genetic testing is a relatively new diagnostic tool. It is still very limited in its scope and expensive to perform. In this chapter, we could only touch on the most pressing issues. We also could only briefly address the wider social context that genetic testing will bring with it in the future. But it should be clear that the seemingly unstoppable march of the new genetics is giving rise to profound social, legal, and ethical dilemmas. The new genetics will undoubtedly touch every one of us sooner than later. Jackie and Emma were aware of their vulnerability but chose to act in a manner aimed at saving their lives. One day, it may be common knowledge to us that all of us harbour genetic mutations that could impact negatively upon our future health. Regardless of that happening, our genetic data ought to remain private, now and in the future. Those data belong to us. They are a significant part of who we are.

Although genes have a significant role in life, they alone do not determine our health, personality, nor future. They act in concert with many other factors, ranging from the combined interaction of all genes in an organism to interaction with the internal and external environment.. Similarly, genes alone do not determine our social well being. That is shaped by human society. Some groups within society have recognised the immense power that comes from prying into an individual's genetic make-up. But their interest has not remained unnoticed. More and more voices are demanding legislation to protect us from these 'gene-hunters', whose actions potentially threaten discrimination in employment and insurance.

The more we arm ourselves with knowledge about the issues of genetic testing and privacy, the better we can respond to any developments we consider not to be in the interest of humanity. For medicine and for us as individuals, the Human Genome Project promises tremendous benefits. Let us not throw away those promises by allowing the abuse of genetic information about our basic self.

9

Knowing your genes: who will have the last laugh?

ROSALEEN LOVE

The public pronouncements that are made about the gene—the gene for this illness, for that aspect of human behaviour—seem to make sense to us. They are given scientific credibility; they are reported favourably in the media (as White points up in chapter 2). They come out of a popular theory about our genetic heritage. Ever since genetics first emerged as a science, it has been widely assumed that genes must influence not only the structure and function, but also the behaviour, of living things. If it is genetic organisation which permits the life of the cell, then genes must be significant determinants of many distinctively human attributes. Or so it has been imagined. Hence headlines in the news: the discovery of 'the' gene for alcoholism, obesity, or Alzheimer's disease (*The Australian* 15 October 1997); there are even notions that there are genes which might promote adultery, be responsible for criminal behaviour (Gans 1997) or be linked to high intellectual ability (Connor 1997); and there is the overwhelming endorsement of the inevitability of the new science—'Genetics—the future is now'.

These headlines indicate the gene has not stayed snugly in its 'objective' scientific place. Now a cultural icon, it has helped to produce increasingly powerful images influencing the way people think and speak about themselves. Myers (1990) argues that the gene may be understood as a code or system of meanings. This chapter explores representations of the gene in the contemporary public culture of Australia, highlighting the so-called 'gene for death' and jokes about genes.

By focusing upon the 'gene for death' I hope to capture the paradox of the gene: that while life arises from genes, those genes also contain the potential for both deformity and death. Yet, while nothing in life is more inevitable than death, humans manage to exhibit a remarkable resilience to the notion, with humour a popular way of coping. Equally, in the face of a dominant scientific message about genetic determinism—that we are somehow victims of our biology—there has been a noticeable social resistance through jokes about genes. Jokes about genes indicate that there is a whole other informal debate happening in the public culture in reaction to what might be termed the 'nothing-but-ism' of the dominant scientific discourse, that is, the notion that we are nothing but our genes.

GENETIC LITERACY

Scientific literacy is a term coined to express the idea that people living in a scientific culture should have some basic understanding of the scientific principles underpinning that culture. It parallels the view that everyone living in a literate culture should be educated to read and write. Two reasons are usually advanced for why scientific literacy is desirable. First, it will help people understand and hence accept the necessity for technological change (Hazen and Trefil 1990). The second is the more radical proposal that scientific literacy empowers people to assume an active role in shaping technological change for socially useful ends (Wynne 1991).

A new phrase—'genetic literacy'—has emerged as a way of conceptualising the duties of the modern citizen in a time of rapid biotechnological change involving significant advances in the knowledge of genetics. Soon, I shall need to know my genes in order to be able to deal with the range of decisions I must make, or assessments which may be made of me (Turney 1993). These decisions may range from the personal level of genetic screening for family planning and *in vitro* fertilisation, to chromosomal monitoring and genetic screening in the workplace, to DNA fingerprinting in the criminal justice system. I shall need to make 'genetically-informed' decisions about whether I want to eat certain foods, or undergo gene therapies. Clearly, then, I will need to know my genes.

If we think of the gene as something fixed—as a fixed unit delivered through time to new generations—we can trace a path of inheritance. I look at my hands and I see my mother's hands. I imagine this as a gift from her, given to me as genes. I imagine she, in turn, received this gift from her parents. I look at my feet, and I see my father's feet. I imagine this gift from him, from his parents to him, and so on back through the generations. I imagine the segment of DNA, the gene, as something fixed, which is handed along as a gift. From where did this gift arrive? How far back must we travel? Initially I imagine human characteristics, but I soon learn this is human-centred, anthropocentric, of me. Scientists say we share something like 99 per cent of our DNA with the primates, 90 per cent in common with the rat. The bacteria, virus, and the common rat have something in common with my genome.

In the history of genetic diseases, the gene is, again, presented as a fixed unit of inheritance. Genetic diseases are particularly awful in the transmission from unknowing parent(s) to a child. Genes for Huntingdon's disease, Tay-Sachs disease and other diseases which kill the recipient, horribly, either late or early in life are examples here. In reflecting on the history of disease, I contrast genetic disease, the often fatal internal flaw, with epidemic disease, the bacterium, the virus, which strikes from without. Yet the two are linked. Humans have survived attacks by viruses and bacteria. But they have survived not just by conquering the infection, 'throwing it off', as we say, but by incorporating part of the invading organism *within* the human genome. One of the fascinating insights from recent work in immunology and genetics is that gene fragments from organisms that caused previous devastating plagues like the Black Death have been found to be incorporated into the genetic make-up of modern humans (Lederberg 1991).

Miroslav Holub, the Czech immunologist and poet, has long reflected on the poetic implications of his scientific work. The poet in Holub (1993) marvels at the genome, 'the logical record of an integrated organism's inner evolutionary drama'. I may start with imagining my genes as a gift from my parents, but I find I possess a genetic gift from all life on earth. Holub sees the inner evolutionary drama played out in the genome. The genome is, he argues, using the notion of DNA as inscription in the Book of Life, 'a genetic chronicle, a good fifth of which is written in absolutely primitive, viral syntax'.

I now imagine the human genome as something of a conquistador, opportunistic, entrepreneurial, a take-over merchant in the battle against microorganisms which treat humans as their prey. Some swift deals have been made in the genetic past. My respect increases. 'It is no co-incidence', Miroslav Holub continues, 'that some of our (relatively noble and relatively human) factors and intercellular signals have nucleotide sequences very similar to retroviruses'. Former predators lie tamed within.

I began by imagining the gene as a fixed unit of inheritance but have found there is a certain amount of fluidity in the notion of 'fixity'. To this I must add, as I explore the notion of genetic literacy, the concept of mutation. The gene may be subject to change, either as the result of an error in the process of replication, or as a result of environmental factors. Mutation is defined as any replicable change in DNA sequence. A good gene changes to bad, replicates and magnifies the error. Genes for particular kinds of cancer are being talked about in two ways; the gene may be a 'bad gene' in that it is inherited as a gene for a particular form of cancer (breast cancer and colon cancer figure largely in these narratives); or it can be a 'good gene gone bad', a gene inherited as harmless, but which suffers a mutation and becomes cancer producing (*Time* 25 April 1994, p. 48). The good gene goes bad and takes control of its immediate environment. Cells multiply, soon out of control. Malignant tumours grow because cells, which normally have a death switch, cannot turn that death switch on—they grow out of control, and the sufferer often dies (Marx 1993).

THE GENE FOR DEATH

Each day it seems as if a new gene for this or that disease is being proclaimed in terms which reinforce the sense of genetic destiny. People speak out about difficult personal decisions made in response to the news that they bear a 'bad' gene. Social workers become genetic counsellors. Different groups have entered the debate about genetic issues. For example, the Central Australian Aboriginal Congress refused to participate in the human gene collection activities of the Human Genome Diversity Project—redefined as the 'Vampire Programme' by indigenous peoples; the Federal Privacy Commissioner has inquired into protection for genetic information collected in medical tests and criminal cases; the Commonwealth Department of Environment hosted a round table on 'Access to Australia's Genetic Resources' (*Gene File* 1994), which was still unresolved at the time of writing. Yet what 'gene' is it that is so often discussed in these various contexts? The gene is rarely given a scientific definition in these public debates. Rather, it is a term to which participants bring diverse and everyday interpretations of meaning. In the near future, however, the phrase 'knowing your genes' will take on a new urgency with the production and consumption of 'gene-test products'.

'Gene tests. How you will die' were the large words on the street publicity-flyer of the *Australian* on 8 August 1994. The article's caption 'Genetic Tests put Fate up for Sale' (p. 3) revealed that knowledge of how you will die will soon be available for purchase in the marketplace as gene-test commercial products. Tests for cancer, heart disease, asthma and senility are said to be on the way, following on the success of a gene test for cystic fibrosis now going into commercial production as a mouthwash. 'Knowing your genes' means knowing how you—or your children—will die. For example, with the cystic fibrosis test potential parents can determine what chance their future children have of inheriting a double dose of a recessive gene for what is termed 'the most commonly inherited killer disease'. Socially, the genetic destiny transition is from 'how you will die' to 'how through abortion you may prevent a killer disease in others', highlighting that gene-testing technology does not lead to any 'easy' solutions (as Rowland also discusses in chapter 6). Issues of choice, ethics, social responsibility and moral beliefs are opened up—all of which are beyond the 'fixity' of genes.

The paradox between public pronouncements of the notion of genetic destiny and the social reality of treatment has often led to a noticeable use of disclaimers from scientists. Take the comment from Dean Hamer (now, Chief of Gene Structure and Regulation research at the US National Cancer Institute) in publicity for his discovery of the 'gay gene' (reported in the 1994 ABC-TV *Four Corners* programme); 'many people have the idea that genes are destiny and that genes are some sort of master-puppeteers that are jerking our strings, sexual and otherwise . . . that's a very incorrect view of how genes act'. This disclaimer came after a number of images had seemingly pointed in the opposite direction, that is, towards genetic destiny. Two gay brothers were shown giving blood for the project. They looked like brothers. The marker pen pointed to bands on strips of x-ray film—the bands from the gay brothers appeared to coincide. The narrator said 'we think that part of the [X] chromosome contains a gene that influences their sexual behaviour'. The marker pen pointed to the spot. The viewer saw it. Both brothers have it. The message is, genes are destiny, even if—as is often the case—a disclaimer accompanied the story.

Again, what is understood from media reports of Hamer's research is often more than is said. Reports are interpreted as the discovery of the 'gay gene', the 'gene for homosexuality'; what is actually said on the Hamer programme is something less than this. Reference is made to 'this portion of the X-chromosome which the two brothers have in common', and 'we think that part of the X-chromosome contains a gene that influences their behaviour', but the message seemingly delivered is that 'the gene for homosexuality' has been discovered (Maddox 1993). 'Knowing your genes', for a gay person, now means what? Comfort for some; for others, indignant rejection of assertion of genetic determinants for behaviour freely chosen. American biologist Garland Allen suggests that instead of talking about *the gene for* some factor, the notion of *the gene that has a norm of reaction for* a particular factor should be substituted. This takes into account the complex relationship between the gene and the environment in which it may, or may not, be expressed—a complex relationship to which Dean Hamer was alluding in his comment above.

Saying the gene is destiny helps define the issue in a certain way, a way that may be perceived quite positively. 'Designer destinies' is the title of an article in the *Economist* (1994), subtitled: 'the first steps towards human genetic engineering give no cause for alarm'. To see genes in terms of 'designer destinies' acknowledges and promotes the notion that there may be potential economic benefits from gene therapy. The issue of 'death and the gene' is redefined in a way that promises some hope rather than no hope. 'Knowing your genes', in this case, brings new choices.

Dissociating the issue of genetic destiny from negative themes is only possible up to a point. The biggest negative theme of all for humans is, clearly, death. What is new about recent pronouncements about the gene and cancer is the degree of certainty of the statements coming out of cytobiology and cytogenetics. Information about 'the gene for' this or that kind of cancer is being given in much more sharply focused, much more dogmatic form. If you have this gene, the headlines report, you will be 99 per cent certain to contract this form of cancer. It will be 'virtually inescapable' (*Australian*, 19 March 1994).

Particular families are profiled. *Time* (in its Australian as well as its US edition) printed a photograph of American Anna Fisher, which showed her surrounded by photographs of her mother and other female relatives who have all died from breast cancer. 'Malignancy is simply part of her pedigree', the article comments. Fisher recovered from ovarian cancer. Then she was given, and took, the option of a prophylactic double mastectomy (*Time* 1994, p. 26); that is, she did not have any signs of breast cancer at this time, but she chose to avoid any chance of getting it in the future by having her breasts removed. With the pictures of those who had died, as well as the stark—supposedly 'rational'—choice made by the survivor, this material is both evocative and extremely powerful.

'The Big Killers', a table in the article, listed estimates from the American Cancer Society about thirteen forms of cancer and their mortality and survival rates, and risk factors. Such a list is, however, much more than summary information simply transmitted to the reader. Against the items on this list the reader may place the name of a person: my mother, my cousin, my friend, and potentially, myself. In this scenario science becomes, as Carey observes, 'a world of dramatic action in which the reader joins, a world of contending forces, as an observer at a play' (Carey 1989, p. 20). But the underlying tension is that we are not only observers but, being mortal, are also participants.

Does knowledge of which genes might kill us provide new freedom, new options? Genetic therapies are not yet available, and this is one of the problems with the release of information about 'bad' genes. But now women like Anna Fisher have the freedom to make a grim choice of prophylactic double mastectomy. It is a freedom to choose, rather than freedom that is granted when restrictions (for example, genetic restrictions) are removed. The restriction, the bad gene, is still there. Anna Fisher knew it in a way her mother did not. She had a consumer choice. She had the knowledge to help her make a hard decision—and made it. The term, 'the gene', imposes order on biological existence. It grants a new form of identity, a genetic identity to the individual. It helps inform narratives of self, providing a sense of something solid and basic, however unwelcome, to the understanding of 'how you will die'. 'I am bound, yet I am free'. The conditions of my being bound, the new genetics and the scope of my genetic literacy, are also the conditions of my new freedom.

GENETIC LITERACY AND THE FLYING PINK PIG

The term 'genetic literacy' embraces a number of contrasting perspectives. One perspective is unashamedly science-centred—relating to what scientists do to convince the public to believe in, and cooperate with, the process of (bio)technological change (Hazen and Trefil 1990). This perspective works according to a one-way transmission model of communication. In this model, science produces a certain field of knowledge and then informs the public, which the public then takes and acts upon to make 'informed' choices or so-called 'rational' decisions.

An example of science-centred communication was the CSIRO travelling exhibition on genetic engineering that toured Westfield shopping malls of Australia in 1992–93. The exhibit was crowned with a series of flying pink pigs—porco-avian icons aimed to ridicule critics of biotechnology. 'Of course pigs will not fly as a result of the new biotechnology' was an explicit message from the exhibition, intended to reassure those with doubts about genetic engineering, and to convince others about its attractions (an aspect Hindmarsh also remarks upon in chapter 3). The CSIRO team sought to instil genetic literacy and provide reassurance with a mix of facts, words, personalities, concepts, history, and ethical implications. Interactive skip videos allowed a level of public participation. At the end of each video sequence viewers were asked to record their vote on a controversial issue, such as 'who should control genetic engineering?'. The tally of votes was then displayed, allowing participants to see where their vote fitted in the public response (Love 1993).

A second perspective on genetic literacy arises in resistance to the science-centred stance. Brian Wynne suggests that scientific communication, if it is uncritical in its praise, may send an unintended message to the receiver that science is above human scepticism (Wynne 1991). A typical response may be 'well they would say that, wouldn't they?'. Or, as I suggest later, if a dominating sense of genetic essentialism is conveyed, resistance might occur in the form of jokes about genes.

We can also ask other questions: what might genetic literacy look like, twenty years into the future? What will genetics mean by then? Where will people turn for genetic information and advice? What will motivate them to do so? Some of the answers might well alarm the scientists. What if a New Age public gives as much scientific legitimacy to the so-called 'intuitive sciences'—palm-reading, astrology, clairvoyance—as to genetic engineering? What if, instead of official biomedical science, alternative genetic therapies are sought? Genetic literacy might then become something very different from TV viewers trying to interpret the meaning of signs on DNA marker gel to workshops on getting in tune with your DNA.

How popular culture has constructed and pushed a certain image of the gene was recently tackled by Dorothy Nelkin and M. Susan Lindee in *The DNA Mystique: The Gene as Cultural Icon*. Nelkin and Lindee collected examples of conversation about genes from sources in US supermarket tabloids, soap operas, parenting manuals, biographies of Elvis, and more. They argue that the images and narratives of the gene in US popular culture reflect and convey a message of genetic essentialism, the notion that the gene marks the essence of human identity. They see these notions as dangerous, lending themselves all too easily to misuse in the service of socially-destructive ends (Nelkin and Lindee 1995). Within the world of genetic essentialism, 'knowing your genes' is tantamount to knowing your biological constraints and thereby accepting a certain biologically-allocated place in society.

JOKES ABOUT GENES

Notions of genetic essentialism as a dominant cultural notion has met (expectably) with resistance. One form of resistance is found in satire. Jokes about genes position themselves against the prevailing genetic essentialism. People telling jokes are taking a sceptical look at the messages of science mediated through popular culture. They are saying, 'Oh yes, and what else?' or, more cynically, 'what a load of rubbish'. In response to being told one's place, told to accept what others deem good for us, one reaction is to poke fun at the new technology and those promoting it. Jokes about genes indicate a certain scepticism especially directed to the institution of medical science, and news broadcasts which report new miracles of modern genetics in the context of global problems, crises in health care, poverty, famine, and war.

In one episode of *Kittson and Fahey* (ABC-TV, 1 November 1993) Mary Anne Fahey and Jean Kittson are shown puffing away on cigarettes. The scene is a factory shed, framed in a smoky haze. Mary Anne confides to Jean that she cannot help it. She has the gene for smoking. In another episode the mad scientist turns mad genetic engineer. Jean tells Mary Anne about the doctor Mary Anne is about to consult: '[He] is into gene shearing, impregnating women with the genes of goats, sheep, pigs. Breeding a woman that could cook, clean, and bring home the bacon too'. Mary Anne is deeply worried. In the 1993 stage show, *Look at me when I'm talking to you*, Dame Edna Everage claims she is going to have a baby, but not in the usual way. It is growing in a saucer on the doctor's window ledge. It all dates back to when Dame Edna was staying with her doctor, and found in the fridge a phial with a use-by date. The name on the phial was Arnold Schwarzenegger, and the contents a product of Arnie's early days as a struggling young sperm donor 'before the germinator became the terminator'. Dame Edna has fond hopes for her offspring: 'It will have my body, my brains and his income'.

The beauty of the throwaway line is that it may cleverly compress pages of logical argument into one telling, funny phrase that will remain in the mind for a long time. Masters of this trade are the comedians H.G. Nelson and (Rampaging) Roy Slaven. Roy is cheerfully omniscient, always ready to comment on any topic from life on Mars ('Call *that* life on Mars?') to the medical status of a footballer's groin. To celebrate the twentieth anniversary of the death of Elvis, Roy and H.G. produced some Elvis memorabilia for their audience. Roy brandished a toothpick Elvis had allegedly used. From this toothpick, Roy claimed he could extract Elvis's mitochondrial DNA. H.G. Nelson was suitably impressed. 'Clone Elvis?' he asked. 'Clone Elvis? Clone his clothes? Clone his guitars?' Roy nodded. H.G. continued: 'Which Elvis would you clone? Young Elvis?' 'Of course not', said Roy. 'I'd clone big fat Elvis' (*Club Buggery* ABC-TV August 16 1997). The viewer is left with an awesome vision of what the world would be like, awash with cloned fat Elvises.

Jokes about genes provide a bridge between popular and scientific cultures. Games are played with scientific ideas, with both scientists and non-scientists mocking scientific hype. Jokes about genes rise above the charge often levelled at satire, that it uses mockery in defence of a conservative *status quo*. Claims made about the genetic basis of what it is to be human are recognised as very important. But if the message is that the causes of major diseases and behavioural problems are 'in the genes', and if the claim is then made that fixing the genes will fix the problem, sceptics might be expected to respond with jokes. The jokes point to the gap between grandiose claims that science has the answers and the everyday world of risk and hazard where scientific certitude is being increasingly questioned (see Beck 1995).

Again, laughter enters through the gaps in the wall of facts that science has proudly built. What Nelkin and Lindee demonstrate so convincingly in their book is that the gene exists as much as a cultural creation, a cultural icon, as scientific entity. In his narrative of adoption Robert Dessaix said of himself, when he found his biological mother: 'I realised that all sorts of things I had carefully crafted and constructed in myself were actually given to me' (Dessaix 1994). This gift is increasingly being labelled 'the gene'. The gene takes off and takes on a range of cultural meanings that the hardworking laboratory geneticist must blanch to consider. Dessaix reports meeting a room full of his new-found relatives, and finding both he and they shared a common quirk: 'I thought I jiggled my knees because I was stressed or nervous. It turned out everyone jiggles their knees. From here to the horizon they are all knee jigglers'. The humorous observation, and the joke, take off precisely because there is no one meaning we all agree to give the term, 'the gene'. There is paradox and inconsistency in the information that we are given. Learning to live with ambiguity is part of the process of getting to know your genes.

CONCLUSION: THE HUMAN GENOME AND THE HUMAN SPIRIT

I referred earlier to the words of Miroslav Holub, the Czech immunologist and poet. When asked how he saw the relation of science to religion, Holub replied:

> I couldn't say whether I am religious. I would obviously be as a unit, as a poor unit which is just an epiphenomenon of something bigger. The something bigger I would rather describe as a genome and not as a spirit. But anyway, it's something supra-individual; the genetic process of the planet. We are obviously in the position of religious individuals because it is way above our heads and we are not the aim of the process (Holub 1992, p. 20).

Holub goes beyond the here-and-now of his individual life and sees it as part of an evolutionary whole which stretches from the past history of life on earth into its future. He is talking about the relationship of part to the whole of life, not, he stresses, in spiritual terms, but 'as a genome rather than a spirit', or 'a sort of instinct for survival—a deep homeostasis in human life' (Holub 1992). The nineteenth century notion of the Life Force is demystified to the Life Thing (or as the selfish gene)—the entity DNA. Yet, as biologists displace themselves further from Nature, Holub yearns for something more, to somehow experience the relation of the part to the whole as best he can in his life work. He places himself before the genome and the genetic process of the planet in a spirit of humility: 'we are not the aim of the process'. This something bigger, 'the genome', is interpreted as an entity with which scientists may cooperate in extending human understanding.

Holub reads the parts of the DNA we share in common with other mammals as 'the logical record of an integrated organism's inner evolutionary drama'. In the battle between life and death, a compromise has been reached in which life continues by appropriating its former ıies to its own ends. In interpreting the history of life on earth in terms of the conquistadorial activities of genomes, Holub finds acceptance of the fact that the death of one individual is often necessary for the life of others: 'Tragic death is a precondition of biological optimism' (Holub 1990a). He imagines a biological or genetic supraconsciousness to which the question of individual death is not central. The biologist must find meaning in accepting responsibility for the planet as a whole and as a viable system (Holub 1990a). Viability thus takes on planetary dimensions and includes the preservation—not only of human civilisation—but of life in general. Holub questions whether humans have the wisdom to recognise this.

Holub's writings concur with warnings from ecologists and others who have studied environmental degradation. For example, Wes Jackson believes that the sustainable agriculturalist must begin with the idea that agriculture cannot be understood on its own terms. Agriculture arises from nature. He poses the question: Should a crop plant be regarded as the property of the human or as a relative of wild things?

> If it is viewed primarily as the property of the human, then it is almost wide open for the kind of manipulation molecular geneticists are good at. If, on the other hand, it is viewed as a property of nature primarily, as a relative of wild things, then we acknowledge that most of this evolution occurred in an evolutionary context, in a nature that was of a design not of our making (Jackson 1984, p. 122).

Both Jackson and Holub urge us to remember this last point—that evolutionary patterns in nature are of a design not of our making. Yet, scientists are daily assuming otherwise, acting as if genetically-engineered crop plants, or the cloned sheep 'Dolly', were wholly the property of humans. Holub, in articulating a notion of some kind of supra-individual process of the planet, is going beyond the here-and-now of biological knowledge, but doing so in a spirit of humility. While remaining first and foremost an immunologist who values his work because it is squarely concerned with the problem of alleviating human suffering, Holub takes us beyond the materialistic assumption that the gene is merely another object to be manipulated. For Holub, knowing our genes means also acknowledging our ignorance about our genes.

It must be very exciting for scientists working in the new genetics. So much is happening, and so quickly. As David Suzuki has discussed so vividly in the Introduction, research students just out of their postgraduate degrees find themselves in research teams doing manipulations undreamed of when their professors were young. Information is 'tumbling off the walls', as one excited scientist remarked. Because there is so little place for ethics and the social and broader ecological context in science education, the tendency is to apprentice the young scientist to explore science with a model of genetic determinism which seems to work well at the practical experimental level. What these genetic manipulations mean to the everyday lives of people however is a topic remote from the laboratory, and when it does come up, it usually does so in the context of science-as-progress—we will all ultimately benefit from genetic engineering. It is the unintended consequences of these well-intended human actions that are the great unknown in this story. The comic throwaway line, 'Clone Elvis?' raises a spectre of delight for some, pure horror for others. One thing is certain. Twenty years from now—if the recent case of the headless, cloned, frog embryo is anything to go by (see chapter 1)—there will be genetic applications no one outside the lab has yet imagined. They will affect everyone's lives. People may be joking now but tomorrow the joke might be on us. That is why this 'knowing your genes' and the social context of genetic literacy is so important. Not only can such knowledge improve one's ability to resist manipulation, but it can also enhance public debate and participatory decision making in this most uncertain area of technological change.

10
Biotechnology, risk and sociocultural (dis)order

STEPHEN CROOK

This chapter offers a sociological analysis of the 'riskiness' of biotechnology, of the ways in which its proponents and opponents articulate that riskiness, and the ways in which riskiness is managed. It is important to establish what such an analysis can, and cannot, achieve. It most obviously *cannot* provide an expert technical assessment of, say, the likelihood that the pesticide resistance of genetically-engineered crops will encourage the emergence of pesticide-resistant weedy relatives. A sociological analysis *can* focus attention on the ways assessments about such matters are made and communicated to the public. This issue is important because risks of any kind confront us *as* risks only to the extent that we are aware of them. So, while the effects of tobacco smoke on the human body were the same in the 1920s as now, they were *limited* to the human body and mediated to society only in 'hidden' effects on morbidity and mortality rates. Unrecognised, unassessed and unregulated, they did not exist as 'risks'.

It will be argued that the most significant site of risk-management in the field of biotechnology is the site of a struggle over what might be best termed its 'cultural riskiness'. For opponents, biotechnology is a source of novel and monstrous hazards linked to a fundamental shift in the human relationship to nature. For proponents, biotechnological innovations are modest, natural and productive improvements, the latest in a long line of benevolent human interventions in horticulture, animal husbandry, medicine and other fields.

The argument is developed in five sections. The first outlines the concept of a 'risk society' and its relevance to biotechnology. The second explains why biotechnology is 'culturally risky'. The third discusses the links between the scientific control of nature and regimes of risk management. The fourth reviews the rhetorical strategies through which proponents of biotechnology manage its cultural riskiness. The concluding section asks whether the battle for the rhetorical risk management of biotechnology has already been won by its proponents.

LIFE IN A 'RISK SOCIETY'

The German sociologist Ulrich Beck (1992) claims that the advanced societies have passed beyond the condition of 'industrial society' to that of 'risk society'. The structuring principle of society is no longer the production and distribution of goods but the production and distribution of 'bads'—chiefly environmental and technological hazards arising from advanced industrial capitalism. These new risks are ubiquitous and frequently they 'escape perception' (Beck 1992, p. 23): we cannot taste the pesticides on our fruit and vegetables, or see the pollutants in acid rain. However, their consequences are increasingly visible, as in large scale toxic accidents such as Bhopal or in the death of the German forests.

British social theorist Anthony Giddens (1991) has similar ideas about risk. He argues that the present stage of social development imposes upon individuals unparalleled demands to make choices and to monitor their own behaviour, but in circumstances where they are radically 'disembedded' from the institutions—family, community, class—that once furnished trustworthy recipes for the conduct of life. In the place of older social institutions have grown complex 'abstract systems' that order our lives in the advanced societies. Examples include money and the financial system, arrangements for medical care, the generation and distribution of electrical power, and the networks through which food is produced, packaged and distributed.

The generally smooth operation of these systems reduces the likelihood of adverse outcomes, but raises the stakes on system-failures when they do occur. So, contemporary mass production and distribution systems for food may be more hygienic than the localised practices of the past, but the consequences of a single failure—contamination of a batch of processed meat or peanut butter, say—are much higher. These low probability but high consequence risks are part of a 'sombre side' of modernisation (along with ecological damage, totalitarianism and war). They promote a 'risk climate' in which risk anxieties and assessments become general and constant preoccupations (Giddens 1991, pp. 122–4).

There are direct connections between these arguments and current anxieties about biotechnology. For example, fears about the invisibility of extensive and potentially high-consequence risks underpin debates about the labelling of genetically modified food. Two British supermarket chains (Asda and Iceland) have banned unlabelled modified foods.

Iceland is concerned 'that a government decision to allow the import of soya, some of which is genetically modified, means that it could be used in processed food stuffs without being traced' (Neale 1997). A molecular biologist is quoted as saying that genetically modified foods 'could give rise to the unwitting spread of new poisons, new allergies and even a reduction in a food's nutritional value' (see Neale 1997).

Similar themes emerge in Australian coverage of the same 'gene bean'—the Monsanto product tolerant to the broad-spectrum herbicide Roundup (another Monsanto product). The policy officer of the Australian Consumers Association protested that the bean had arrived 'willy nilly' with other soya beans and could find its way into some 60 per cent of supermarket foods (Hoy 1996a). The general concern that lies behind these specific complaints has been formulated by a British 'prize winning microbiologist' in terms that are very close to those used by Beck: 'Introducing a gene into another organism is a Russian roulette process—the position the new gene occupies is not controllable. We are being asked to partake in a nutritional experiment of global proportions' (Uhlig 1996).

Proponents of biotechnology are irritated and frustrated by such claims. The response of the Australian Food Council to the 'gene bean' controversy is typical: the beans constitute 'no public health or safety hazard whatsoever' labelling would be 'inappropriate, impractical and meaningless' the gene bean has 'the same composition, nutrition and processing characteristics as conventional soybeans' (see Hoy 1996a). British government scientists have complained that 'ill-informed "scaremongering"' about genetically modified foods has produced unwarranted consumer resistance in Britain (Brown 1997). That story illustrates a point about biotechnology disputes made by Levidow (1996, p. 55): both sides can play the 'risk' game. Proponents of biotechnology allege scaremongering could 'cost Britain the international race to produce healthier foods, fuel and "greener" industrial materials'. A failure to develop agricultural biotechnology would be globally risky because 'high-performance genetically modified crops have a vital role in feeding the world's population which is expected to double over the next 40 years' (Brown 1997).

THE CULTURAL DYNAMICS OF RISK

In the Beck–Giddens analysis 'risk' is definitively modern. For Beck (1992, p. 55) risk is 'a systematic way of dealing with hazards and insecurities induced and introduced by modernity itself'. Giddens (1991, pp. 109–11) also insists that a modern orientation to the future, in which hazards become risks, must be distinguished from traditional orientations to fate and destiny. Such sharp distinctions between the traditional and the modern have been contested by the American anthropologist Mary Douglas. For Douglas, first, all societies develop mechanisms for processing perceived hazards and second, all risks are culturally selected for attention. These two claims establish a close connection between risk and sociocultural order: in any given society, 'to alter risk selection and risk perception . . . would depend upon changing the social organisation' (Douglas and Wildavsky 1982, p. 9).

Douglas helps to show why biotechnology is 'culturally risky', a threat to sociocultural order. One of her interests has been the way in which cultures define themselves in relation to risks of pollution (see Douglas 1966). Witchcraft, wrongly prepared food, menstruating women, and particular animal species have all been 'selected' by some culture or another as a fundamental threat to its 'purity' and sociocultural order. Such threats are both made visible and controlled in the institutionalised injunctions, prohibitions and ritual observances through which cultures defend themselves. Douglas (1983) has argued that Western anxieties about environmental pollution should be understood as an example of this syndrome.

Pollution threats arise when boundaries that define a sociocultural order are breached: when non-food species are eaten, when the milk is mixed with the meat, when humans meet with spirits. This conception of pollution connects with the principle that sociocultural order is built from basic binary oppositions: nature–culture, male–female, raw–cooked. Biotechnology is a culturally risky threat to order because it breaches boundaries between 'natural' and 'cultural/artificial' that have served as ordering principles in Western cultures. Campaigns against the use of r-DNA-derived milk-boosting growth hormone somatotropin (bST) in milk production are surely opposing this kind of cultural 'pollution' when they emphasise the natural 'purity' of milk and the threat of 'contamination' posed by bST (see Hannigan 1995, p. 169).

The very principle of the genetic modification of plants and animals breaches established distinctions between the natural and the artificial. Non-modern societies often use myths and folk-tales to explain the origins and consequences of pollution: the marriage of Coyote to Yellow Corn Girl brings witchcraft to the Pueblo, Pandora opens the box and releases disease and pain into the world. Modern societies, too have their representations of the dangers of mixing the natural and the artificial, from *Frankenstein* to *Terminator* and *Jurassic Park*. To the British press, genetically modified crops are 'Frankenstein food'. The cultural riskiness of biotechnology, understood in this way, is at the heart of debates about biotechnological risks and will be considered in more detail shortly.

CONTROL, RISK AND RISK MANAGEMENT

Different societies 'select' different risks for attention and these selections are closely connected to sociocultural order. We would expect a society built around fear of invasion to be different from a society built around fear of famine, and different again from a society built around fear of witchcraft. Of course, matters are never that simple: multiple risks must be confronted at the same time. But it remains true that the measures a society puts in place for the identification, assessment, and management of risks—its 'regimes of risk management'—play a critical role in defining an overall sociocultural order. The generic 'cultural riskiness' of biotechnology is amplified by the absence of a single authoritative regime of risk management in societies such as Australia. Rather, three regimes compete with, overlap and undermine each other. This claim cannot be argued in detail here (see Crook 1997), but it can be approached through the relationship of biotechnology to the broader scientific management of natural risks through enhanced control over natural processes.

Modern science has been characterised as a long-term project aimed at human control over nature. Its success has been linked to the 'reductionism' of science—its search for the basic component parts of natural phenomena. Once these parts are identified and the relations between them specified, larger-scale processes can be controlled through the manipulation of the parts. As Charlesworth *et al.* (1989) have argued, a plausible account along these lines can be given of the emergence of biotechnology out of the 'new biology' of the 1940s. What had been a 'third world scientific territory' (Charlesworth *et al.* 1989, p. 39) given to observation, classification, and speculation was transformed by the migration of physicists such as Schroedinger and Slizard. Under their influence, molecular biology emerged as 'the investigation of the ultimate units of the living cell in the same way that physicists and chemists investigate the ultimate units of matter' (Charlesworth *et al.* 1989, p. 42; see also Wills discussion in chapter 5). This mode of investigation is taken to be linked to an inherent 'logic' of control, so that it is a short step from Crick and Watson's 1953 model of the 'double helix' to genetic engineering. As Gottweis (1995, p. 202) puts it, 'the goal of engineering life was . . . inscribed into molecular biology from its inception'.

As a project for the control of nature, science relates to risk management in three main ways. First, science and technology provide tools for the direct, first-order, management of natural hazards. From scientific medicine to power generation to weather forecasting, science and technology are at the heart of many 'expert systems'. Second, the ways in which risks are identified and assessed have themselves become increasingly 'scientised'. For example, the forms of risk insurance—life, medical, automobile—which we take for granted could not have become established until the mathematics of normal distribution and statistical sampling were systematically applied (see Bernstein 1996). Third, as Beck argues, many environmental risks are now 'second-order' natural risks. For example, a new medical treatment, a nuclear power plant or a genetically-modified organism may have hazardous consequences that are worse than the hazards the innovation is designed to control. To ensure that this is not the case, experts conduct risk assessments and trials before the medical treatment is recommended, the power plant built, or the transgenic organism released.

An important consequence flows from these connections between science and risk. The whole complex task of risk identification, assessment and management can appear as a purely technical matter, best left to the experts. Returning to the example of the Monsanto 'gene bean', who is to assess the risk posed by this product of the plant geneticist's craft but other plant geneticists? Hence, the exasperation of the Australian Food Council in the face of complaints about the unlabelled release of the bean: the competent experts have declared the bean safe—identical to other beans—so objections can only be ill-informed, irrational or mischievous. Irritation at the ignorance and irrationality of those who cannot grasp the logic of risk assessment is widespread among risk specialists and promoters of new technologies.

Governments responsible for the regulation and management of risk find themselves in a quandary, torn between public disquiet and expert advice. An important current of US opinion holds that risk management has been too 'public-opinion based' rather than 'risk based', and that this must change. As Zechauser and Viscusi (1996, p. 14) put it 'government policy should not mirror citizens' irrationalities but . . . should make the decisions people would make if they understood risks correctly'. At the other end of the spectrum stands Beck's cynical view of expert risk assessment: because science bases risk assessment on case-by-case, laboratory-based, experiments and can never hope to match the complex interactions and synergies of the 'real world', such scientific assessment is limited and ultimately fraudulent. Science has become 'the protector of global contamination of people and nature . . . the sciences have squandered until further notice their historic reputation for rationality' (Beck 1992, p. 70).

'Pure' calculations of risk, uncontaminated by culture, are not possible: there is an irreducible sociocultural dimension to risk. More specifically, the identification, assessment and management of risk can take place only within patterned policy settings and institutional arrangements that can be termed 'regimes of risk management'. The regulation and eradication of risk has been a major element in the programmes of what Beck (1996) has termed the 'provident state', and others the modern 'welfare state'. With the support of major interest groups, such states put in train a regime of 'organised' risk management. Legislators, experts and enforcement agencies cooperate in the systematic identification, assessment, and management of risks—ranging from industrial pollution and infectious disease to unemployment and illiteracy.

Over the past twenty years or so the bureaucratised and centralised structures of the 'provident state' have come under attack, so that Australia is now well advanced down the road of a 'neo-liberal' redefinition of the role of the state. The symptoms are familiar and include: sales of public assets, financial and trade de-regulation, labour market reform, the 'downsizing' of the public service. In risk management there is a corresponding scaling-down of the organised regime. In areas from emergency services to air traffic control and meat inspection there is a trend towards industry self-regulation and the 'contracting out' of services. Perhaps most critically, there is a heavy emphasis on individual responsibility. As Petersen puts it, 'neo-liberalism calls upon the individual to enter into the process of their own self-governance through processes of endless self-care, self-examination and self-improvement' (Petersen 1996, pp. 48–9). In the emergent 'neo-liberal regime' of risk management, state services are transformed into bureaux for the provision of information and expert advice to 'responsible' individuals and industries.

In Australia, the European Union and the United States, biotechnology is subject to the bureaucratic surveillance and regulation associated with the organised regime. Arrangements for its regulation often seem *ad hoc* and complex, the result of uncertainty about where biotechnology 'fits' in pre-existing bureaucratic structures. Consider the arrangements for the assessment of genetically modified food crops. In Australia, responsibility falls to the Genetic Manipulation Advisory Committee (GMAC) with its four sub-committees (for its historical context see chapter 3). GMAC is responsible to the Department of Industry, Science and Tourism, but its advice goes to a number of other Commonwealth departments, notably Health and Environment. State governments are also involved. In Britain, two government departments and seven advisory committees share responsibility under over-arching European Union regulations. In the US the Department of Agriculture, the Food and Drugs Administration (FDA), the Environment Protection Agency and the National Institute of Health have a stake in biotechnology regulation at the federal level. The individual States also have wide regulatory powers.

This pattern of management of biotechnological risks gives rise to two contrasting criticisms. On one side, the industry complains of over-regulation. The Australian Biotechnology Association (1996a) complains that regulatory complexities 'can cause difficulties and delays and increase costs'. On the other side are fears that current regulatory regimes cannot grapple with the broader and long-term risks of biotechnology. In an echo of Beck's thesis, the (British) Nuffield Council on Bioethics has expressed concern that in case-by-case regulation 'no one is considering the long term implications of releasing [genetically altered] plants into the environment'. The council's secretary has claimed that this concern is shared by many regulators (Irwin 1997). Gottweiss (1995, p. 217) similarly argues that German regulators defined biotechnological risks on a case-by-case 'additive' model, excluding 'the complex interactions of "downstream" causation'.

There is presently little evidence of 'neo-liberal' risk management in biotechnology, but it will surely develop. For example, if the labelling of genetically-modified foods becomes entrenched, a degree of responsibility for the management of the risks of such foods will be shifted to the consumer in the neo-liberal manner. Again, if the genetic screening of human populations becomes established (as both Tonti-Filippini and Gesche suggest in this volume), it is easy to imagine that parents will be made responsible for decisions affecting the genetic health of their offspring. This possibility is canvassed by Beck-Gernsheim (1996) and in Wilkie's (1996) dystopia where parents agonise that they cannot afford to buy the top-of-the-range IQ for their child from 'Genes R Us'.

There are peculiarities to the management of biotechnological risks that seem to fit neither the 'organised' nor the 'neo-liberal' regime. Consider three propositions. First, American research has shown that 'the implementation of governmental controls over risk appears to reduce, rather than to increase, people's personal sense of powerfulness' (Priest 1995, p. 47). Second, Zinnen (1994) cites Mark Cantley of the OECD as arguing that specific provisions for the regulation of biotechnology 'are intended to protect biotechnology from the public, rather than . . . the public from biotechnology'. But such reassurance-regulations fail because they send a mixed message: government statements say 'biotechnology poses no special risks', while government actions in establishing special regulations say 'biotechnology poses special risks'. Third, while the Australian Biotechnology Association (1996a) is concerned about the complexity of the regulatory regime, it is cautious about 'simplified and uniform legislation' covering genetically-modified organisms on the grounds that 'there appears to be no valid reason for treating these products any differently from the same products made by conventional means'.

These peculiarities suggest that there is a ritualistic and rhetorical dimension to the regulation of biotechnology. At issue is not the quantifiable risk of a specific adverse outcome, but rather a general public unease at the cultural riskiness of biotechnology. Both the implementation of standardised procedures for risk assessment and the language used by promoters of biotechnology can be analysed as elements in a third 'ritualised regime' of risk management that offers reassurance in the face of general unease. The repeated formulae of risk assessment and biotechnology promotion both neutralise cultural riskiness.

RITUAL AND RHETORIC IN BIOTECHNOLOGY DEBATES

Biotechnology is 'culturally risky' because it threatens sociocultural order by breaching boundaries that have ordered Western culture. This riskiness can be managed if the boundaries can be redrawn, or if biotechnology can be shown not to breach them. The strategy of redrawing the boundaries connects with what some commentators see as the rhetorical work at the scientific core of biotechnology. Charlesworth *et al.* (1989, p. 41) note that the 'new biology' is distinguished by its use of metaphors of 'information' and 'codes' which transform 'the phenomena of life processes into something that can be *read*'. Gottweis (1995, p. 197) pursues the point more critically: the capacity of biotechnology to 'rewrite life' makes it 'a new technology of power, with the potential to transform fundamentally the society that is to benefit from its advances'. Gottweis's pivotal point is that the genetic 'rewriting' of life allows human progress to be redefined in terms of the potential interventions of biotechnology. On this basis, the cultural risks of biotechnology can be managed by 'the inscription of deficiency into life' (Gottweis 1995, p. 199).

Levidov (1996, p. 60) develops a similar argument, noting that agricultural biotechnology represents its activities as the rectification of 'genetic deficiencies' in crops. Biotechnology offers modest improvements on an occasionally deficient nature, giving a helping hand to evolutionary processes (see Levidov 1996, p. 62). These rhetorical moves do not take place in a cultural vacuum. The biotechnological conceit of 'raw' nature as dirty and deficient finds many echoes in the era of what Clark terms the 'over-exposing' of nature. In scenic routes and forest walks through national parks and in the simulations of the tourist resort, zoo or theme park we are well-used to processes in which 'the surface of the natural environment [is] reworked into a more pristine and more visible version of itself' (Clark 1997, p. 84). Similarly, Smith (1996, p. 37) charges the eco-friendly US chain 'The Nature Company' with 'the simultaneous idolisation and commodification of nature combined with . . . effacement of any distinction between real and made natures'. In this way we become 'softened up' for the idea of an 'improvement' of nature that is itself natural.

The rhetorical battle over the cultural riskiness of biotechnology is fought along two main axes, one running between 'natural' and 'unnatural', the other between 'old' and 'new'. If 'new' and 'unnatural' are both risky, it is important to its proponents that biotechnology should not be seen as having both characteristics at once. The ideal, but perhaps implausible, strategy would be to position biotechnology as both 'old' and 'natural'. Failing that, it might do to have its 'unnaturalness' tempered by age, or its novelty tempered by 'naturalness'.

The 'rewriting of life' is a strong and prevalent version of the claim that biotechnology is 'natural'. Hannigan (1995, p. 170) reports that the claim—'that biotechnology is merely an extension of nature and consequently can be expected to be perfectly safe'—forms one of three main planks of Monsanto campaigns. There are many other examples of this move. The Australian Biotechnology Association (1996a) argues that 'evolution itself results from successful mutations that occur randomly in nature. The new techniques of biotechnology simply increase the rate and precision of such changes'. A US FDA-sponsored 'bST Fact Sheet' from Cornell University handles the issue with some finesse. It is stated that bST is 'normally' produced in the pituitary gland of dairy cows; it is one of 'a group of hormones produced naturally in the cow'. Nothing is mentioned about the method of production of r-bST, and the expanded acronym is used only twice, the favoured term being 'supplemental bST' (Barbano 1997).

The European Federation of Biotechnology (EFB)—through its Website (1994)—links the issue of 'naturalness' with that of 'novelty'. It is acknowledged that people doubt 'the safety of products involving genetic modification since it is seen as an "unnatural" process'. However, science sees things differently: the distinction between 'natural' and 'genetically modified' is 'a distinction not made by scientists'. To show how the lay distinction is inappropriate, the EFB (1994) points out that: 'almost all common foods in our diet come from the breeding, hybridisation and selection of plants, animals and microorganisms over many centuries. These are also genetic techniques which could therefore be regarded as "artificial"'. At the very beginning of the paper the EFB emphasises the long history of 'biotechnology' by using different adjectives—but the same noun for—its 'traditional' (wine, cheese and beer making) and 'modern' (r-DNA and cell fusion) forms.

This view of biotechnology as the up-to-date version of the wholesome cottage industries of the past is endorsed by the Australian Biotechnology Association (1996b), for whom the main difference between 'modern biotechnology' and 'traditional methods' in stock breeding is that the former produces 'hardier and more productive stock more quickly' than the latter. The same connection and contrast between the traditional and the modern can be found among the boosterish stories that form the staple diet of Australian press coverage of biotechnology. The link with tradition serves to ease any anxieties about the cultural riskiness of 'modern' biotechnology, while the break with tradition emphasises the superior efficiency of the modern. A *Sydney Morning Herald* story (Hoy 1996b) links a project seeking to establish a sheep-milk cheese industry at Cowra and making use of genetic mapping techniques to select stock, to the Rocquefort cheese mould, and to Julius Caesar and Charlemagne.

That there must be a contrast, as well as a link, between the modern and the traditional is critical to the promotion of biotechnology. The modern must be like the traditional, only better. Another *Sydney Morning Herald* story (Hoy 1996c) reports an Australian robot—Vitron 501—that can clone three million trees a year. It will accelerate the environmentally desirable shift from logging old growth forests to plantations, produce export income and replace 'tedious' manual work that has a high risk of repetitive strain injury (RSI). Seedlings for cloning are identified through 'DNA fingerprinting' that allows selection after eight weeks rather than the traditional 30 years.

To summarise, the cultural riskiness of biotechnology is managed 'ritualistically' through 're-assurance regulations' and through a series of rhetorical tropes that neutralise one or both of its dangerous characteristics—novelty and artificiality. The vigorous promotion of the benefits of biotechnology is also a tool for repositioning its riskiness: if there are a few risks, they are surely worth running. Biotechnology will feed the hungry, cure the sick, restore the environment and reinvigorate the economy (see Hannigan 1995; Levidow 1996). Those who fail to see this are either ignorant or mistaken. And the critics who would stem the biotechnological tide are viewed as placing in jeopardy human progress.

THE RHETORICAL TRIUMPH OF BIOTECHNOLOGY?

In a study of debates in the British House of Lords on embryo research, Mulkay (1993) argues that participants deployed two distinct 'rhetorics', where a rhetoric is 'an interrelated set of background assumptions plus typical assertions that are evident in participants' discourse on a particular topic' (Mulkay 1993, p. 723). The 'rhetoric of hope' is 'part of the taken for granted, dominant discourse of science in our society'. At its heart is the 'interpretive link between research, control and human benefit' and it seems 'natural' to most people most of the time to think of science as progressive and beneficial. By contrast, the second 'rhetoric of fear . . . is part of a culturally subordinate discourse of science articulates an image of 'moral decline and of socially disruptive changes brought about by a scientific community increasingly out of control' (see Mulkay 1993, p. 726–38).

Boosterism about biotechnology resonates with the 'rhetoric of hope'—which is the dominant discourse on science (in chapter 2, White relates how this is replayed in the print media). By contrast, critics of biotechnology appear to have at their disposal the much inferior resource of a subordinate and marginalised 'rhetoric of fear', the most substantial bearers of which may be horror movies of the 'mad scientist' variety (Mulkay 1993, p. 737). On this basis, and on the evidence of the skill with which the cultural riskiness of biotechnology is neutralised, it may seem that the war over the cultural meaning of biotechnology has been won by its proponents. Thus, Hannigan (1995, pp. 175–6) can assemble a list of reasons why r-bST is unlikely to become a public issue on the scale of other environmental controversies: there is little intra-scientific opposition, it does not resonate with global 'social justice' issues, opponents have little social power, and media interest is lukewarm. In Hannigan's view, the best weapons the opponents of r-bST have are rhetorical (the 'purity' of milk). If this is so, their campaigns might seem doomed, given the formidable rhetorical resources at the disposal of the proponents of r-bST and other biotechnological artefacts.

But rhetorical battles are rarely won outright, and there are flaws in the strategy of the proponents of biotechnology that may at least postpone their final victory First, the biotechnological 'reinscription' of life is not without problems. As Nelkin (1996, p. 24) argues, geneticists have elevated the genome to the status of 'the eternal and fundamental basis of human identity'. The writing of DNA becomes a 'sacred text', the decoding of which enables the geneticist 'to reconstruct the essence of human beings, unlocking the key to human nature'. This 'genetic essentialism' plays into the hands of Christian opponents of genetic engineering, especially when it involves human genes. Nelkin quotes from a fundamentalist journal: 'is it permissible to alter humanness at its core, to tamper with our essential humanity ?' (Nelkin 1996, p. 23).

Second, media and industry-style boosterism may be counter-productive. For example, Priest (1995, p. 45) argues that one-sided coverage is 'more likely to exacerbate than to calm public concerns both because of the boomerang potential of one-sided messages and because lay publics bring a definite agenda of concerns to the task of understanding news about biotechnology'. A letter to *The Sunday Age* (17 August 1997) illustrates both of these points:

> . . . whenever I read an article about genetic engineering . . . a feeling of profound uneasiness comes over me but, like so many, I don't know

> enough about the technology to be able to argue the case as to why it might be dangerous. I find it quite extraordinary that such radical changes are being made to what constitutes the foundation of our existence with so little public debate.

The letter shows, first, that 'genetic essentialism' can promote anxieties about biotechnology beyond the Christian Right. The author accepts that genetic engineering touches the 'foundation of our existence' but takes that to be a ground for concern. It shows, second, that 'reassurance strategies' discussed earlier in this chapter (and see Zinnen, 1994; as well as chapter 3) can be counter-productive.

Public anxieties are profound—but unfocused—because they derive from biotechnology's cultural riskiness. They can be neutralised in ritualised practices and rhetorics which will be effective if they and their sources are trusted by their intended audiences. But those reassuring practices and rhetorics are likely to be *distrusted* precisely because they are recognised as aiming only to reassure. There may be a curious convergence between the interests of the promoters and opponents of biotechnology here. Both may be best served by a more open and public debate than has taken place in Australia to date. For opponents, such a debate would provide an opportunity to place better their case on the public agenda. For promoters, an open debate would give an opportunity to establish levels of public trust that would lend credibility to their attempts to 'reassure'.

Part 3

Molecular farming, novel foods and campaigning

11

Gene-biotechnology, the state and Australia's agri-food industries

GEOFFREY LAWRENCE AND FRANK VANCLAY

Biotechnologies are considered by scientists, government officials and rural producers, to be the latest in a long line of 'high tech' inventions which will bring great advantage to Australian agriculture and food manufacturing.

In previous decades, Australian farmers utilised the latest tractors, headers, ploughs, fertilisers, pesticides—and of course, improved seed varieties and animal breeds—in an attempt to reap profits and to remain internationally competitive. Similarly, food processing companies have been continually looking for ways to enhance their market share by developing new products, and by reducing costs through innovative technologies.

For conventional farmers, new inputs allow for improvements in labour efficiency and for increases in plant and animal productivity. Improved plants may produce more grain, use less fertiliser, be harvested in a shorter time, or resist plant pathogens. Improved stock may convert pasture grasses more efficiently, tolerate ticks or drought, reach slaughter age more quickly, or better suit consumer markets. Conventional breeding programmes have enabled scientists to make small but important improvements to crops, pastures and animals. But greater improvements are deemed to be necessary in the competitive national and international markets to which farmers sell produce. Farmers face circumstances often involving overproduction (which forces prices down), so-called 'unfair' barriers to free trade such as tariffs and import restrictions, subsidisation of farmers by overseas governments which undercuts prices received by Australian farmers, and the overall tendency for farm input costs to rise faster than the rate of returns for farm output (see Vanclay and Lawrence 1995; Malcolm *et al.* 1996). Their challenge is to produce more, and more efficiently, and to do so in ways which enhance sustainability of the environment.

The food industry faces many of the same pressures as the farmers. It has to compete with overseas products often 'dumped' on international markets, and must find means of ensuring that foods exported remain appealing to consumers in distant locations. This means preventing spoilage, increasing shelf life, and enhancing flavour—all within a competitive pricing regime. It is here that genetically-engineered plants, animals and manufactured foodstuffs are being touted as revolutionary new ways to attain efficiency gains in farming and in food processing *simultaneously* with improvements in environmental security.

AGRICULTURAL BIOTECHNOLOGY IN AUSTRALIA

Leaving aside 'basic' and human health research areas, the application of biotechnology to agriculture and the food industry (or agro-biotechnological research) fits within seven broad categories: animal health; resistance of plants and animals to diseases; pests and environmental extremes; the production efficiency of crops, pastures and animals; production efficiency in food processing; the quality of food and agricultural products; development of new products and processes (such as novel foods and biological pesticides); and the environment (Lee 1992a).

Recombinant-DNA research currently conducted both in Australia and abroad includes: the cloning and breeding of animals (for example, transgenic animals, growth hormones); vaccine production for the sheep, beef and pig industries; the development of insect-resistant plants through the insertion of *Bacillus thuringiensis* (Bt) genes; the improvement of cellulose digestion in ruminants; the use of hormones to produce low fat meat (particularly in pigs); and gene-modification of plants and animals to allow them to produce complex proteins for use by the pharmaceutical industry.

Some projects have special relevance for Australia. CSIRO scientists are using r-DNA methods to alter forage plants so that they might produce more cysteine—a scarce amino acid. Sheep ingesting the altered plant species will be able to produce more wool. Experiments are also being conducted to improve the ability of ruminants to digest cellulose. Given Australia's arid landscape, it is hoped that if digestibility of forage increases, so will the output of animals in marginal (as well as in the more fertile) regions.

In a similar experiment, scientists are attempting to alter rumen bacteria so they can detoxify the plant poison, fluroacetate. If the research has a positive outcome, graziers in northern Queensland will overcome current stock-loss problems. CSIRO researchers have spliced into a cotton plant the gene conferring resistance to the broad-spectrum herbicide 2,4-D. CSIRO scientists have also successfully transferred a gene (the common French bean to the field pea) which produces a protein causing starvation in insects. When insects begin to eat the pea crop they 'feel' full, but actually starve to death. Plant geneticists are also attempting to develop sugar cane plants that can tolerate frost, flowers which remain fresh for longer periods, salad vegetables which don't 'brown', bananas which contain doses of vaccines which prevent gastro-diseases in children, and novel plants including the blue rose. Animal geneticists are seeking to make sheep resistant to blowflies, and cattle to temperature and water stress.

During 1993, a patent was taken out on the so-called 'super pig'. It was the first patent to be taken out on an animal in Australia and is one of few such patents worldwide. The 'super pig' carries about ten extra pig growth hormone genes. These genes—switched on in the presence of zinc—are constructed from a small segment of synthetic DNA copied from human DNA material. The transgenic animal has very lean meat and grows some 20 per cent faster than normal—and utilises less food than a conventional pig (see *Sydney Morning Herald* 1 April 1993, p. 3).

Recombinant-DNA technology is viewed as advantaging the environment through: the production of biological or 'natural' pesticides (such as Bt) which reduce the need for the application of increasingly potent agrochemicals; engineering plants and animals which need less water—viewed as highly desirable in a land where water is a scarce resource; and controlling pests like foxes and rabbits through induced sterility (see Lee 1992a).

In the food industry, the basic thrust is toward the conversion of agricultural products, by-products and substances currently wasted in food processing into value-added materials (see Britz 1991). Recombinant-DNA techniques are being employed to replace chemical additives in foods and to enhance colour, texture and other properties. Activity to date has been in improving starter cultures in dairy products, making cheese production more efficient through r-DNA derived rennet, developing 'natural' preservatives such as bacteriocins (which eliminate specific spoilage organisms), developing quality assurance tools, and improving the nutritional value of foodstuffs. Nationally, one of the main aims of public research is to discover ways of using agricultural surpluses via the conversion of bulk products into new substances (see Healy 1991).

These examples indicate the general approach of Australian research and highlight the typical output-boosting, production-efficiency focus of much of the work. By increasing plant and animal resistance to disease, improving plant and animal efficiency and their capacity to survive in arid Australian conditions, research will allow producers to continue along the 'productivist' path (see Lawrence 1995). The concern is that, rather than providing an alternative to the current 'high tech' approach to rural production—one which has brought with it environmental pollution and destruction on a massive scale—it will reinforce the current trajectory toward specialisation, intensification and, more generally, toward ever-increasing surpluses and further environmental destruction (see Sleigh 1988; Bunyard 1996).

Many ecologists, social scientists and public interest groups question the ability of companies, such as Monsanto, which are major players in biotechnological development—but whose profits are currently based upon the sale of environmentally-damaging chemicals—to provide long-term solutions to these problems. It is argued that genetically-engineered products will be tapered to the existing chemical strategies thereby conforming to, rather than challenging and overcoming, agriculture's chemical 'fix'. As Kloppenburg (1991, pp. 488–9) once mused: 'I would like to think that—properly used—biotechnology can provide us with important tools to move towards a more sustainable agriculture. But can we trust Monsanto to get us there?' Some posit, instead, the creation of 'superweeds' and 'superbugs' which have become resistant to the genetically-engineered seeds which are packaged for the farmer, along with proprietary chemicals, and sold by agribusiness (see Kloppenburg and Burrows 1996; Steinbrecher 1996).

STATE SUPPORT FOR GENE-BIOTECHNOLOGIES

With its history of publicly-funded research in agriculture aimed at improving plant and animal productivity, Australia has reacted to the brave new world of genetic engineering by wholeheartedly embracing it. It has paid for its scientists to go abroad to study the latest techniques, has conducted major studies and held important conferences, has identified biotechnology as one of Australia's main research foci, and has put in place a number of measures to stimulate the development of the bioindustry in Australia—including a National Biotechnology Programme.

From the 1980s, the state has sought ways of promoting private capital investment and involvement in Australian agro-biotechnological development. In the CSIRO, the direction of research has moved from traditional means of breeding toward genetic engineering and staff have been told that their promotional opportunities will improve if they develop industry linkages (see Lawrence 1987). This was in concert with two other developments—a commercial arm called SIROTECH was established to assist scientists to patent their inventions and to contract the development of products to outside agencies; and, divisional chiefs were directed to obtain at least 30 per cent of their departmental funding from 'outside' (generally private) sources (see Lawrence 1987; Lee 1992a).

Many of these changes had a major impact upon the work of Australian scientists. With funding cuts occurring in particular areas of public research, new collaborative arrangements with private firms were entered into. There were also changes to the management of research.

At the beginning of the 1990s a House of Representatives Standing Committee was charged with the task of reviewing the development and potential release into the environment of genetically engineered organisms (GEOs). In arguably the most important report to the government on biotechnology the Committee recommended, among other things:

- that existing guidelines on experimentation and release be tightened (specifically, that scientists be forced, in their grant applications, to set out 'worst case scenarios' in the case of escape or eventual release of genetically modified (or engineered) organisms (GMOs/GEOs);[1]
- that federal and State governments better co-ordinate their monitoring of health and safety standards relating to r-DNA experimentation;
- that patents on GEOs be extended beyond the conventional sixteen years (so as to allow companies time to recover investment costs); and
- that mandatory approval from a new GMO Release Authority be obtained for release of all GEOs (including products containing live GEOs) (Lee 1992a).

In effect, the Committee acknowledged some potential problems with existing regulations. Its recommendations however were largely designed to allay the fears of those community groups which had expressed major problems about release of GEOs (see Hindmarsh and Hulsman 1992). It nevertheless viewed gene-biotechnology as having considerable economic and other benefits and whose development should be monitored but not hindered (Lee 1992a; see also chapter 3 for an analysis of the inquiry).

Other government-commissioned reports highlighted the value of biotechnology in assisting Australia to achieve economic growth—including its integration with the Asia–Pacific region. A federal Department of Industry, Technology and Commerce report suggested that:

> there needs to be closer links between breeders, researchers, food processors and exporters so that efforts can be better focused and guidance given on those areas which give the best economic returns relative to the research and development investment made. In this way Australian producers will be better able to meet the international needs of food producers and will be able to provide Australian processors

> access to better raw materials for the export of processed foods. Otherwise, improved application of biotechnology by Australia's international competitors could ensure that Australia will remain an exporter of bulk commodities (Invetech 1989, p. 67).

In another report prepared for the Bureau of Rural Resources (see Healy 1991), eleven strategic technologies were identified as assisting agricultural industries to diversify, to add value, and to improve sustainability of natural resources. Genetic engineering was placed first. Yet, according to a group of some fifty leading scientists interviewed in the study, there was evidence that Australia's research base in genetic engineering was eroding (Healy 1991, p. 17). So, against the general tendency for the state to withdraw from supporting the agricultural sector, the report instead argued that the state should invest in agrobiotechnology in order to promote productivity improvements in farming and allow food processing industries to 'value add' (see ASTEC 1982). Significantly, while funding for biotechnology research was viewed as crucial, state funds to support alternative approaches to production—such as organic farming—were nowhere to be found.

The recent urgency of the 'productivity push' in Australian agriculture is also directly relevant to Australia's balance of payments and debt problems. Large sections of Australian farming are facing severe economic and social problems (see Burch *et al.* 1996; Lawrence *et al.* 1996; Gray *et al.* 1997). Strategic technologies such as biotechnology are viewed as essential to improve the economic position of farmers and to address Australia's trade imbalance (see Bureau of Rural Resources 1991; Healy 1991; Vanclay and Lawrence 1995).

GENE-BIOTECHNOLOGY AND AUSTRALIA'S INTEGRATION INTO THE GLOBAL FOOD ORDER

A most enthusiastic voice for biodevelopment as a key avenue for Australia's integration into the Asia–Pacific region has been that of Australia's peak establishment farming organisation—the National Farmers' Federation (NFF). The NFF endorsed the OECD's prediction that without biotechnology, increases in agricultural productivity would soon flatten (see NFF 1993). Moreover, according to the NFF (1993, p. 84): 'Trends in rural production already suggest that future differences between the output of developed and developing countries will be based on their respective level of biotechnology use'. Biotechnology is viewed as being vital to expanding the sales of Australian primary products—and especially in the role of helping to feed a burgeoning Asian population (see NFF 1996).

In the early 1990s, industry analysts argued that biotechnology would become a 'strategic' technology in the processed food industry. On the basis of overseas trends, key developments were seen to be: the improvement of enzymes through genetic engineering; the detection of food contaminants; the production of flavours, fragrances and sweeteners; and the further development of fermentation processes (Scott-Kemmis and Darling 1991, p. 33). Complete endorsement of a biotechnology future for Australian foods came from the corporate-linked Australian Food Council (1996). Opportunities were seen to exist for food processing firms to form backward linkages to agriculture, and for chemical and biotechnology firms to become involved in food processing. Many of the largest and most prominent of the food manufacturing firms operating in Australia use modern biotechnology. Such firms include Arnott's, Biotech Australia (AgrEvo), Bunge, Burns Philp, Carlton and United Brewers, Kraft General Foods, Mauri (Gist-Brocades), United Milk Tasmania, Solvay Biosciences, Pfizer, Christian Hansen, Novo, APV Baker, and Alfa Laval (see *Australasian Biotechnology* 1991).

Promoting itself as the 'clean' food nation, Australia is expected to be well placed to capture new market share in the unprocessed and processed foods sectors in the Asia–Pacific region (Prime Minister's Science Council 1991; East Asia Analytical Unit 1994; Lester 1994). To achieve this, the inevitability of a biotechnology strategy was outlined by government:

> It is essential to have public acceptance of recombinant DNA technology if we are to move to have commercially viable industries in agriculture and in biotechnology. It must be stressed that it is not just the biotechnology industry *per se* and its export potential that is at stake but the survival and competitiveness of our major industries in agriculture, both plant and animals that are also at stake. Major sections of the rest of the world are moving into using genetic modification to improve the efficiency of agriculture and we cannot afford to be left behind with outdated technology (Bureau of Rural Resources 1991, p. 61).

Here, then, is an explicit belief that biotechnology represents the future. In fact, six developments have converged to promote biotechnology—and particularly genetic engineering—as a key component for an agriculture and food industry of the new millennium. These are: the need to boost farmers' productivity in a more highly competitive global trading arena; scientific advances in biotechnology; the state's demand that publicly-funded science be more closely linked to private industry; the growth in importance of large trading firms—especially in relation to their connections with Asia; new consumer markets in the Asia–Pacific region; and, public demands that farming systems become more sustainable and that environmental degradation and pollution be overcome.

The question is: will the application of biotechnology fulfil current expectations?

THE UNCERTAIN PROMISE

The issue of whether farmers will benefit from the new biotechnological inputs is of great importance. If agribusiness produces and distributes the new genetically-modified seeds and livestock, there is no guarantee that rural producers will benefit. These inputs will, after all, be the exclusive (private) property of the companies developing them. Today, input prices are rising much faster than returns for farm goods: costly biotechnological products might be expected to reinforce this tendency. Some question the likelihood of benefits flowing to farmers as a group, and note the current polarisation in agriculture, claiming that an agribusiness future will be an industrialised agriculture with fewer, more wealthy, farms among a great many smaller, economically marginal, producers (Kloppenburg 1988; Geisler and Lyson 1993; but also see discussion in Krimsky and Wrubel 1996). For example, herbicide-tolerant plant species are viewed not as freeing agriculture from chemicals—one of the promises of biotechnology—but of ensuring the dependence of farmers on proprietary brands. This has been referred to in colourful terms as 'bioserfdom' (see RAFI 1997). If herbicides increase their presence, there is the real possibility of chemical resistance occurring in weeds, and of chemical pollution in waterways—again, the opposite of the biotechnological promise (see Kloppenburg and Burrows 1996).

With respect to cleaning up the environment and reducing the amount of land required for production (see Lee 1992a), it is necessary to question the possibility for either outcome in Australia. While some biotechnologies may have the potential to eliminate introduced feral pests (through biotechnologically-induced sterility), to clean up oil spills, or to reduce the use of some herbicides and pesticides, the introduction of GEOs in the environment may well result in the transfer of genes between species (including the imparting of resistance in undesirable bacteria and plants), increase the risk of escape of pathogenic organisms, and disrupt and potentially damage biotic communities (see Hindmarsh 1992b; Steinbrecher 1996). As Yearley (1996, p. 42) suggests: 'engineered (species) cannot easily be both widely used and closely contained. There will inevitably be some risk of genetic pollution and in this case it will be pollution by genetic material unprecedented in nature'.

And, while increases in the production efficiency of 'enhanced' plants and animals may, in theory, allow for reductions in land use, it is more likely that farmers will seek to achieve greater volumes from existing lands—as a strategy to offset cost-price pressures through greater levels of output (see Lawrence 1987). In other words, the intensification of agricultural production will continue unhindered, thereby increasing pressures on Australia's fragile environment (see Lawrence *et al.* 1992; Conacher and Conacher 1995). It may also see new transgenic crops and livestock developed for Australia's more marginal inland regions—potentially reeking environmental havoc in those areas.

There is a growing list of concerns regarding release of GEOs (see Tait 1990; Busch *et al.* 1991; Goodman and Redclift 1991; Hindmarsh 1992b; Steinbrecher 1996). Two main issues are of particular note. The first is an ecological one: Australia is particularly vulnerable to 'foreign' organisms and the potential extinction of indigenous species including the incompatibility of GEO release with Australia's desire for biodiversity and sustainable development. The second issue is about food and consumers: is the extension of biotechnology compatible with the trends in 'green consumerism' both in Australia and in food-importing nations? These are discussed below.

Ecological vulnerability and biodiversity

In terms of ecological vulnerability, problems caused by the past release of foreign species designed to improve the profit levels of farmers are poised to occur again. For example, when the South American cane toad (*Bufo marinus*) was introduced via Hawaii to control the cane beetle, it was not foreseen that it would impact on other species, or that it would expand its niche beyond the cane fields. It has now spread out of control in Queensland, the Northern Territory and New South Wales, leading to the death of native animals that prey upon amphibians. It is also outcompeting native frogs. Insufficient research was conducted to establish the likely impact of the toad. The uncontrolled spread of the calicivirus following its accidental (some speculate, deliberate) release on the Australian mainland in 1996—while having an overall beneficial outcome of substantially reducing the wild rabbit population—confirms the inability of scientists to contain their field experiments. The ensuing illegal introduction by farmers of the calicivirus into neighbouring New Zealand is yet another example (see Anderson 1997). Moreover, there is always the possibility of wilful neglect. In July 1997 a scientist at the Queensland University of Technology was suspended for knowingly exposing colleagues to the risk of contracting Japanese encephalitis while he undertook unapproved experiments on the deadly virus (*Campus Review* 23–29 July 1997). Although these three cases are not related to genetic engineering, they nevertheless confirm the public's fears about the difficulties faced in keeping scientific experiments under control.

With strict guidelines and a stringent monitoring system in place, are these sorts of 'accidents' likely to occur in relation to genetically engineered species? In fact, one already has. It occurred in South Australia in 1988 with the release of transgenic pigs. Without obtaining full permission, the firm Metrotec sent a batch of the pigs to Adelaide's abattoir for slaughter. While, in this case, there was virtually no potential for the escape of 'new' genes from the animals, it represented a major breach of government guidelines (see Lee 1992a, pp. 190–3; also see Hindmarsh's discussion in chapter 3). The point is, with so many rural producers wanting immediate productivity gains, and with biotechnology companies wanting a return on their substantial investments, there is inevitable pressure for the release of novel organisms (Hindmarsh *et al.* 1991; Kloppenburg and Burrows 1996). Scientists do not fully understand the ecological complexities of existing systems, and cannot give any certainties that novel genes will not transfer to native species (see Hindmarsh *et al.* 1991; Burch *et al.* 1992).

The issue of foreign genes in the environment is linked strongly to concerns about biodiversity. If new, productive, genetically-altered species are economically more beneficial for producers, they may be adopted *en masse*, reinforcing current patterns of monovarietal production in Australia. Yet, despite short-term productivity gains they might not provide long-term economic or ecological benefits (see Burch *et al.* 1990). The narrowing of the genetic base for agriculture is inconsistent with Australia's avowed desire to embrace ecologically sustainable development (ESD). In its draft national strategy, the ESD Steering Committee (1992) stressed the need for both nature conservation and environmental protection as part of a broader strategy to protect diversity. In this report—and in the earlier document of the Working Group on Sustainable Agriculture (1991)—the complicated issue of biotechnology's role in relation to biodiversity was largely ignored.

It should be noted that various surveys in Australia indicate most rural producers and scientists believe strongly that genetically-engineered plants will reduce pesticide use, are compatible with the nation's 'clean and green' image, and that they will contribute to a more sustainable agriculture (see Lawrence and Norton 1994; Foster and Ghonim 1995). Those largely unconvinced are organic growers (who fear that any transgenic genes in the environment will be a form of 'genetic pollution'—see Lawrence, *et al.* 1993) and consumers (see Norton chapter 13). Women, in particular, appear more concerned than men about environmental and social consequences (see Foster and Ghonim 1995).

Consumers and foods

The issue of a biotechnologically-oriented agricultural and food industry future in a world of 'green consumerism' is yet to be addressed in Australia. Increasingly more discerning consumers—concerned about food safety and nutrition (see Allen 1993)—are eschewing chemically-laden products, purchasing 'clean and green' products, and are demanding the labelling of any genetically-manipulated (or so-called 'foreign') foods (see Harper 1993). Growing evidence from home and abroad shows that consumers are wary of genetically-modified foods (see chapter 13). In one national survey, Australian consumers exhibited a 'latent anxiety' about genetic technology, with four in five respondents believing genetic engineering could create new diseases as microbes escape from laboratories, and with two in three believing genetically modified plants pose a long-term hazard to the environment (see *Australian* 23 May 1995, p. 8). In one small-scale study, consumers were greatly concerned about novel foods: biotechnology for animal health and welfare was one thing, but to feed the family genetically-engineered foods was quite another (see Norton and Lawrence 1996; Norton *et al.* forthcoming 1998). In past years, Japanese, Korean and American governments have been quick to reject any foods which have been 'tainted' with chemical residues. Australian grain farmers have been warned that Japan—one of their premium markets—is likely to be strongly resistant to transgenic crops (Bolt 1997, p. 7).

Biotechnology may result in significant improvements in food output and efficiency, but ironically, also lead to the rejection of those foods by concerned consumers (and by countries which, in trying to assist their own farmers, are seeking ways of rejecting the products of competing nations). Hoban (1989) has suggested that consumer opposition is likely to grow where the public believes food has in some way been 'contaminated' by its manipulation with new biotechnologies. He notes that 'public concern for technological risk is greatest when risks are seen as involuntary, exotic or unfair. These characteristics apply to biotechnology as seen by the public' (Hoban 1989, p. 20). In the US, consumers place more importance on cultural values associated with food (for example, what is wholesome and pure) than they do on scientist's assessment of 'risks' (Busch 1991).

Simply, consumers do not share the same assumptions as scientists about what constitutes natural food. Consumers decide whether a food is desirable or undesirable on the basis of a combination of subjective beliefs and objective understanding. They are becoming more wary of the health dangers of some processed foods, rejecting the 'more quantitative logics' which have been pursued by agriculture and the corporate food system (Marsden 1992, p. 220). There is also evidence that consumers are becoming more suspicious of the ability of governments to ensure food safety (see Hoban *et al.* 1992; see *Australian* 19–20 October 1996, p. 46). In such circumstances it is not hard to envisage an intensification, worldwide, of the rejection of new plants and animals, and particularly new food products, by consumers. There is strong evidence that this has grown throughout the 1990s, and is continuing today. For example, although organic food production currently represents only 1 per cent of the food market, the Geneva-based UN Conference on Trade and Development forecasts a climb to as much as five to 10 per cent in Europe and North America by the year 2000 (*International Agricultural Development*, 1996, p. 23). Increasing consumer interest is being driven by such things as the 'mad cow disease' crisis, supermarket interest in organic products, the growth in outlets of natural food stores, escalating evidence of the environmental impact of intensive farming, and consumer concerns about food quality and any health repercussions.

Yet, for the pro-biotechnology Australian Food Council (AFC), consumers are viewed as being out of touch with the benefits of biotechnology. Apart from those who have 'real moral or religious concerns', concerns about food biotechnologies arise 'from a lack of understanding and knowledge' (AFC 1996, p. 22). Today, moral concerns are probably the most important objection to genetic engineering and its applications (Wagner *et al.* 1997). Public education is seen by the Food Council as 'vital' for 'successful integration of genetically modified foods and food products into everyday life. Scientists are endorsed and praised by the Council (1996, p. 22) for ensuring that 'experimentations in recombinant-DNA techniques' are 'controlled and safe'.

The AFC's rather simplistic trust in science is at the heart of concerns raised by the environmental and consumer movement in Australia. Is this concern misplaced? In terms of the intense and escalating debate worldwide about gene-technology, it is clear that if only *one* experiment went badly wrong or *one* release had unintended health consequences, Australia's international reputation as a reliable supplier of wholesome foods would be forever damaged. Accidents aside, it must be stressed that what biotechnology's proponents appear to ignore—in the contemporary circumstances of the 'greening' of policy and attitudes throughout the world—is the distinct likelihood of consumer rejection of the very genetically-engineered products which are being promoted in Australia for their ability to improve farm productivity and to increase output in the processed-food industry. This signal has, to reiterate, already been strongly indicated with the growing trend to organic foods and with Japan's possible rejection of novel foods.

CONCLUSION

To improve Australia's competitiveness in the global marketplace, Australian governments, farmer groups and food companies have endorsed agro-biotechnologies as the 'best' option for productivity enhancement. They have been selected on the basis of Australia's traditional 'comparative advantage' in the production of farm products, the likelihood of sales growth in the Asian–Pacific market, and other opportunities for commercialisation. They have been premised on the overall desirability of having products which will add to productivity at the same time as they improve agricultural sustainability.

Although there is *prima facie* evidence to suggest that the genetic engineering path will be one leading to increased output and efficiency, the sociocultural aspects of present farm and food production, together with environmental impacts, leave us sceptical about the eventual benefits. First, many biotechnologies will be purchased from large transnationals which—under patents—will be able to control the price of various products. This is not likely to benefit all farmers—especially struggling family farmers. Second, biodiversity might be threatened by the widescale application of the most productive products. Third, given Australia's dependence upon export agriculture, any reluctance on the part of consumers abroad to purchase the new foods may lead to falling demand, further exacerbating Australian agriculture's poor economic performance. Were this to prove correct, Australia's attempts to place biotechnological innovation at the centre of its farming and food industries may, in time, diminish the export prospects for those industries and undermine the very system of farming it is seeking to support.

It would be prudent for Australians to give considered thought to what sort of food production system they want into the next century. If they have concerns about gene-biotechnology, they will have to act quickly—and politically—to stem an already purposeful industry and government-endorsed rush toward a genetically-engineered future.

NOTES

1 It should be noted here that while the report referred to 'genetically modified organisms (GMOs)', we prefer to retain the traditional term 'genetically engineered organisms' (GEOs). Retaining the term GEOs is important because the business of biotechnology also uses the term GMO to encompass the products of (1) fermentation and plant issue biotechnologies and (2) traditional forms of agricultural and industrial biology. This obscures the radical shift that genetically-engineered products represents.

12

Wearing out our genes? The case of transgenic cotton

ANNA SALLEH

In August 1995, the head of CSIRO's Division of Plant Industry, Jim Peacock, suggested to a scientific conference in Newcastle how the problem of world hunger might be solved. In a keynote address to the Australian and New Zealand Association for the Advancement of Science, he said gene technology could bring the 'doubly green revolution' needed for agriculture to feed and clothe the world's growing population in an environmentally sustainable way (Peacock 1995).

A year later, Australia's first genetically-engineered broadacre crop was sown—30 000 hectares of 'Bt cotton' designed to kill its own pests. The new technology promised to eliminate up to 90 per cent of the insecticide sprays normally needed (Fitt *et al.* 1994). Saving the planet aside, however, cotton farmers had other, more immediate, concerns on their minds.

Cotton is one of the most heavily sprayed crops in agriculture. If the new genetically-engineered variety could reduce spraying, it would help stem the tide of accusations that cotton growing was poisoning people and their environment. Not only that, but the technology could provide a timely solution when the arsenal of effective chemical weapons against crop pests was losing its effectiveness.

Some have described Bt cotton as one of the most successful applications of agricultural biotechnology research so far and a major step towards a more environmentally-friendly cotton industry. Others claim it is just another in a long line of quick fixes that characterises a crisis management approach to pests. They insist that it will be 'smart farming' rather than 'smart crops' that will deliver the much sought after bounty of sustainable agriculture.

This chapter assesses Bt cotton's potential for being a 'clean and green' technology. As such, it will explore one of the major risks it brings—that of 'wearing out' our valuable Bt genes.

COTTON'S PESTICIDE TREADMILL

Monoculture cotton is more prone than many other crops to infestation by weeds and a variety of insect pests. The most formidable pests of cotton in Australia are the 'heliothis' caterpillars of *Helicoverpa armigera* and *Helicoverpa punctigera*. Secondary pests include green mirids, mites, aphids and thrips. Australian cotton growers spend up to $100 million a year on pesticides (NRA 1996a)—over one-eighth of Australia's total expenditure on crop protection products (Avcare 1995).

The average crop is sprayed between six and sixteen times per season. Eighty per cent of these sprayings are against heliothis (McLean *et al.* 1997, p. 240). When pests adapt by developing resistance, however, many more sprays are used. For example, in 1974, just before the cotton industry at Kununurra in Western Australia was wiped out due to resistance, farmers were spraying 40 times a year in an attempt to control insect pests (CC 1996, p. 189). *H. armigera* has developed resistance to most of the chemicals that in the past provided effective control.

Resistance aside, cotton growers also have a severe image problem, often portrayed as environmental pariahs. A relatively small percentage of cotton pesticides applied by air reach the target insects. This means that everything from humans and fish to other crops and export cattle may be exposed to these chemicals as well. While the effect of pesticides on the environment and human health is the subject of intense controversy, the negative impact of residues on the highly valuable export beef market is undisputed (for example Passey 1996).

As the insecticide stockpile dwindles—either due to resistance or concerns over implications for the environment, human health or trade—the monoculture cotton industry is in crisis, and is desperate for a new weapon against heliothis. Over the last decade, biotechnology promised a dazzling solution in the form of transgenic Bt cotton.

BENIGN BT: THE PERFECT CANDIDATE FOR THE FIRST SMART CROP

Bt toxin is produced in nature by a soil bacterium called *Bacillus thuringiensis* (Bt). Of 20 000 different strains, the variety *kurstaki* has been of most use to agriculture. Sprays containing Bt *kurstaki* have been used worldwide to kill Lepidoptera caterpillar (including the diamond-back moth and heliothis pests) in vegetables since the 1920s. Bt is extremely valuable to organic farmers as it is both completely 'natural' in origin, and is quite specific to a small range of insects, while other insects—some useful as predators or parasites in biological control—are left unharmed.

Two years after recombinant-DNA technology was first applied to agriculture in 1981, Bt was identified as a perfect candidate to exploit. Yet, when the Cry IA(c) gene from Bt *kurstaki* was first transferred into cotton it did not produce very high levels of the toxin. In 1988, molecular biologists from agrochemical company Monsanto 'edited' the gene to make it more 'plant-like'. This 'construct'—or designer gene, now patented by Monsanto—performed much better than the original (Monsanto 1995). It was transferred by CSIRO into local varieties of cotton, more suited to Australian growing conditions, in order to produce an economically viable crop.

The advantage of transgenic Bt cotton over Bt sprays were said to be many. Whereas Bt toxin in sprays was exposed to UV light which caused it to break down and become ineffective, Bt toxin in transgenic crops was protected inside the plant cells, where it remained active for longer. Another advantage was that the toxin was produced throughout the whole plant, and at all times. This would ensure that caterpillars, no matter where they were on the plant, would be exposed to the toxin at first bite. Sprays, in contrast, often missed parts of the plant (such as the underside of leaves), or were only applied after the pests had already done some economic damage (Krimsky and Wrubel 1996, pp. 55–72). With the advantages of transgenics, however, came a challenge—that of keeping pests from becoming resistant to the Bt toxin in the transgenic cotton, just as they had to all other insecticides used against them.

RESISTANCE TO BT: THE STORY SO FAR

In every insect population, some individuals possess genes which give them greater tolerance to a particular insecticide. While the chemical will kill most of the population, these 'resistant' individuals survive and over time breed up in numbers. When this happens, farmers tend to use higher and more frequent doses of insecticide—each spray inadvertently selecting for the survival of even more resistant insects. Eventually, such a high proportion of the population is resistant that the insecticide becomes effectively useless.

Already, in areas such as Malaysia and Hawaii, following a number of years of intensive use of Bt on vegetable crops, the diamondback moth developed resistance to the same variety of Bt now being used in transgenic cotton. There are signs that similar resistance may already be developing in Queensland. Bt was used sparingly as a spray for over 30 years and it was only when farmers had lost almost all of the synthetic insecticides to resistance, and concerns over pesticide residues increased, that Bt sprays began to be used more frequently (McLean *et al.* 1997, p. 262; Verkerk and Wright 1997). Indeed, the use of Bt sprays in cotton has increased twenty-fold in the last five years (Forrester 1994).

Today, because Bt is so valuable to organic and biological farmers it is used sparingly as an emergency control should other methods fail, and so that resistance has less chance of developing. While some argue over the economic viability of organic farming (even while its market is increasing worldwide as pointed out in chapter 11), its approach of minimising reliance on any one particular tool is advocated by leading scientists in the area of pest management research. This approach is referred to as Integrated Pest Management (IPM), and also happens to be the catch-cry of those who promote transgenic Bt crops.

THE 'PERFECT' PROMISE OF INTEGRATED PEST MANAGEMENT

IPM—and the more recent Integrated Crop Management—spreads the selection pressure for resistance around by employing a number of pest management tools. These can include chemicals, predators, parasites, pathogens, pheromones, baits and attractants, resistant crop varieties, physical and cultural controls (for example, legislation) and biotechnology (Brough *et al.* 1995). Some, such as biological controls, are less likely to lead to resistance than others. Given the right balance of tools, some argue, it may be possible to avoid resistance developing altogether.

Biological controls, which many see as the mainstay of good IPM, require a healthy and diverse population of insects in the field which is not generally possible in monoculture cotton regularly sprayed with broad spectrum chemicals. While such chemicals act to turn the field into a 'biological desert', the specificity of Bt toxin to Lepidoptera caterpillars, on the other hand, makes it very compatible with biological controls and thus IPM.

Transgenic Bt cotton, it was said, could provide a new 'platform' for IPM (Fitt 1994). This vision was held despite the diamondback moth experience, and despite concerns that the constant expression of Bt toxin in transgenic cotton posed even stronger selection pressure for resistance than Bt sprays.

Around the world, IPM practitioners have been applying the 'evolutionary brakes', and removing selection pressure wherever pests have shown signs of resistance to Bt. UK-based IPM researcher Robert Verkerk argues that transgenic cotton's built-in Bt toxin makes it hard to do this. 'You can't very well ask the farmers to pull out the crop half way through a season,' he says (Verkerk 1997).

Bt cotton researchers fully understood this flipside to the new technology which is why they developed a resistance management strategy that some hoped would stave off resistance for 'many decades' (Fitt *et al.* 1994). Other, more recent estimates, however, are that resistance could develop to transgenic Bt cotton in as little as three years (CC 1996, p. 57). These conflicting predictions call for a closer look at the strategy.

THE RESISTANCE MANAGEMENT STRATEGY FOR BT COTTON

No matter how effective an insecticide, a certain number of tolerant individuals will always survive. The higher the dose of insecticide, the fewer the survivors, but the 'tougher' they will be. In any resistance management strategy it is crucial to deal with these lest they breed up into a whole army of immune 'monster bugs'.

When it came to Bt cotton, researchers reasoned that a suitable field of Bt-free plants located near the transgenic crop could provide a safe haven (or refuge) for heliothis to breed up in, without hindrance from Bt. The insects not exposed to Bt would be more 'susceptible' to the toxin, and by mating with any survivors from the transgenic crop could dilute out the undesirable trait of resistance in the next generation. These refuges were central to the resistance management plan.

Researchers estimated a refuge size of 10 to 20 per cent of the crop to produce enough susceptible insects to mate with resistant survivors in the transgenic crop. They also recommended a range of other controls for farmers using Bt cotton—some of these in strategies to manage resistance to previous insecticides. These included, first, cultivation (ploughing) of soil after harvesting to a depth of 10 centimetres, to destroy any highly resistant pupae and their burrows; and, second, monitoring with a view to spraying (with their usual insecticides) should a 'threshold' number of heliothis caterpillars be detected (NRA 1996b).

Bugs in the system: High dose gone wrong

The refuge strategy was designed to work best when there was a '*consistent and high* [my emphasis] expression of the insecticidal protein *throughout* [my emphasis] the plant' (Fitt *et al.* 1994). The idea was that the dose of Bt produced by the transgenic crop would be high enough to kill virtually all heliothis, leaving only a handful of highly-tolerant individuals which could be easily swamped by an army of 'susceptibles' bred up in the refuge. This would minimise damage and reduce the chance of a resistant population developing. This was called a 'high dose' strategy (Monsanto 1995; McLean *et al.* 1997, p. 259).

Yet, far from being the perfect vision of precision described in public relations brochures, transferring genes across nature's boundaries proved instead to be a very hit and miss affair. Consequently, Monsanto 'took what they could get' (Forrester 1997). While the dose was high enough for most of the US heliothis pests (for which Bt cotton was initially developed),[1] Australian species were genetically more tolerant to Bt which meant more would survive leading to a high risk of resistance (CC 1996, p. 73; Forrester 1997). Although researchers tried to compensate for this 'low dose' by recommending larger refuges than those used in the US (to produce more susceptibles), they were still worried (see Forrester 1997).

Bt toxin is a protein and in transgenic cotton, like other proteins, its level in the plant changes according to factors that affect the plant's growth. NSW Agriculture entomologist Neil Forrester assessed the dose produced would be 'just enough' when conditions were good, but the plants would 'not need much of a hiccup to fail' and make resistance 'a real problem to manage' (Forrester 1997). The outcome was that a significant percentage of the Australian 1996–97 Bt crop had to be sprayed more than expected (CRDC 1997; Pyke 1997). Growing conditions were suspected of being partly responsible (McLean *et al.* 1997, p. 240).

Entomologist Rick Roush from Adelaide's Waite Institute, declared that the crop did not produce enough Bt toxin to be 'conducive to resistance management' (CC 1996, p. 59). Despite Monsanto (1995) claims to the contrary, the expression of Bt toxin in Australian transgenic cotton under real world conditions was neither high nor consistent.[2]

Bugs in the system: beyond dose

Queensland University entomologist Myron Zalucki suspected the 'high risk' strategy would probably not work. Apart from the 'low dose' of Bt, he predicted it would be almost impossible to ensure that the susceptible insects from the refuge were around at the same time and place to mate with the resistant ones. If, instead, resistant mated with resistant, super-resistant individuals might be produced in the next generation. Moreover, insects on the Bt cotton grow more slowly than those on a non-Bt crop (Zalucki 1997). Thus, even if flowering of the refuge crop could be synchronised with the Bt crop, susceptible and resistant insects might still not meet. Also researchers were not sure how far moths would travel before mating. In 1996–97 it was not compulsory for refugia to be adjacent to the transgenic Bt crop. Given this, if farmers see refugia as less productive, they may—especially in hard economic times—choose to put them on marginal land possibly too far away from their Bt crops to be effective (Quantum 1996).

Since the refugia requirement is a condition of Monsanto's licence, it is the company's responsibility to ensure compliance. Even though compliance is legally binding, even normally apolitical agricultural researchers have expressed concern that this apparent self-regulation is inappropriate (Quantum 1996). Despite Australia's international reputation for good resistance management, farmers have not always followed the requirements of previous strategies. Those on tight rotational programmes, for example, prefer to directly drill their wheat crop into cotton stubble rather than plough up the field first, so they may be reluctant to cultivate their cotton fields after harvest as required under the current plan (CC 1996, p. 32). The strategy also conflicts with the desire for minimum tillage agriculture. While technology to make tillage more environmentally friendly is reported to be under development, the 1996–97 crop was grown without it.

THE TREADMILL CONTINUES

Monsanto has said it wants to prevent resistance from 'diminishing the effective life span' of its product (Monsanto 1995), and CSIRO scientists working with the cotton industry are hopeful that resistance management will extend Bt's 'useful shelf life' (CC 1996, p. 62). This very language, however, speaks of a disposable technology—with refugia-based resistance management seen as a tool to 'buy time' (Roush 1997a), until the next chemical or transgenic fix comes along. One anxiously-awaited development is a new variety of cotton which has two insecticidal genes 'stacked' into it—to act as a double hurdle to the development of resistance. One such 'pyramided variety' being developed by CSIRO contains two Bt genes—Cry 1A(c) and Cry IIA. Researchers have warned, however, that even this could be of little use if resistance to the single-gene variety develops and knocks out one of the 'hurdles' in advance. As little as three years use of the current variety could lead to this unfortunate situation (McLean *et al.* 1997, p. 270), while the pyramided variety may not be ready for 10 years (CC 1996, p. 62). Some researchers would have preferred Bt cotton not to have been commercially released until a second Bt gene had already been successfully engineered into the plant, but the cotton industry was desperate and 'just couldn't wait'.

An additional concern with using two-gene Bt cotton is that insects exposed to one type of Bt toxin may develop resistance to a second type even without being exposed to it. Such 'cross-resistance' has already been observed in experiments between the very two toxins planned for CSIRO's pyramided variety (Gould *et al.* 1992). Technical problems may require other insecticidal genes to be found for the second resistance hurdle in transgenic cotton (CC 1996, pp. 62–3). While this may provide a more effective hurdle, the problem is that viable insecticidal genes are not easy to come by (Rissler 1997; CC 1996, p. 64). And there are likely to be concerns about the safety of the few candidates which currently exist (CC 1996, p. 63).

Despite evidence that nature is proving more complex than anticipated, genetic engineers are confident their craft will solve all problems. Sociologists of science have critiqued what they call the 'logical positivism' of biotechnology which 'attempts to reduce nature to small, definable pieces, subject to human manipulation'. They observe this has 'become the dominant epistemology often to the exclusion of important other ways of knowing' (Busch *et al.* 1991). Applied to this case study, it appears that biotechnology, rather than sustainable practices as a whole, are 'signified' as the cutting edge of agricultural research.

In adopting this argument, Jane Rissler of the US-based Union of Concerned Scientists says 'If, in the last 40 years, we had spent as much in this country on sustainable agriculture research as we have on chemical research, we would have a lot of the answers' (Quantum 1996). In the meantime, the National Registration Authority has allowed for an increase in acreage of Bt Cotton each season (NRA 1996a). While researchers recommend the single-gene variety should be limited (CC 1996, p. 59), Monsanto's imperative to make up for ten to fifteen years worth of research through sale of its technology can be expected to act as a counterweight to these views (as suggested by Sharp 1992). Other pressures on Monsanto include the competition from other companies which have their own versions of the Bt gene (Quantum 1996).

There is the promise of new classes of synthetic insecticides that can help 'buy time' until two-gene Bt cotton is available, although according to one commentator these insecticides do not have the potential for a sustainable environmentally-friendly industry (Roush 1997a). Ironically, a current argument over the price growers must pay Monsanto for Bt cotton (Landline 1997) may see growers opt for these chemicals instead of the transgenic cotton the company hopes to keep selling (McLean *et al.* 1997, p. 242). It may be that the single-gene Bt cotton itself is simply used to buy time until a new set of chemicals is available. Whichever way it goes, the pressure seems to be on for the treadmill to continue. To add to this, another reported advantage of Bt cotton—that it requires less spraying—is also under challenge.

TO WHAT EXTENT WILL BT COTTON ELIMINATE SPRAYING?

Long before commercial Bt cotton was planted, Australian researchers knew that when the plant stopped growing towards the end of the season, Bt toxin production would drop off, and the cotton plants would be vulnerable to heliothis attack (Fitt 1994; Fitt *et al.* 1994). While entomologists were worried about the increased resistance this would bring, growers who had massive overheads to cover would be more concerned about an uneconomic drop in yield. If they stopped the normal spray regime for heliothis they might also have problems with secondary pests, unaffected by Bt. It was all very well to have a long-term vision of a more diversified approach to pest management free from broad spectrum sprays, but in the short-term, researchers anticipated that farmers would spray Bt cotton two or three times in the season (even if these sprays were likely to be broad spectrum, and thereby defeat the selectivity of Bt cotton by killing beneficial insects).

What researchers did not anticipate was the degree of unpredictability of the transgenic crop in the real world situation. While early estimates were that there would be a 90 per cent drop in sprays, for the 1996–97 Bt crop as a whole there was only a 52 per cent decrease. The decrease was not evenly spread among producers. While some fields required no sprays, others were sprayed just as many times as conventional cotton (CRDC 1997; Forrester and Constable 1997). Yields, too, were variable and with more sprays than they had budgeted for, some farmers complained they had paid a 'Rolls Royce' premium for something that ran more like 'a Holden' (Landline 1997).

A PLATFORM FOR INTEGRATED PEST MANAGEMENT?

While some hope transgenic Bt crops will smooth the way for a broader-based IPM in cotton, the reality may be somewhat different. 'IPM . . . means different things to different people', the Australian Cotton Conference was told in 1996 (CC 1996, p. 153). For example, while organic farmers try to maximise biological controls to minimise the exposure of non-target organisms (such as humans and beneficial insects) to toxic chemicals, the cotton industry has—to a large extent—based its approach on rotating the use of broad spectrum insecticides.

While biological controls such as food sprays are being researched in Australia today for use on cotton, they require further work before they can be considered for widespread use (*Australian Cotton Outlook* 1997, p. 14). And by the time they are ready, it may be too late—at least for Bt. Food sprays—patented by Rhône Poulenc—have been hailed as the perfect biological control complement to Bt cotton with the proviso that 'Bt resistance does not quickly muddy the waters' (CC 1996, p. 76). Food spray researcher Robert Mensah of NSW Agriculture admits that it will be a challenge to complete research before resistance develops to Bt cotton (Mensah 1996).

The food spray system involves the use of strips of lucerne throughout the cotton field to lure green mirids away from the cotton. The lucerne also attracts beneficial insects which can in turn be lured into the cotton crop using a food spray, when the time is right, to prey on heliothis. The food spray also repels heliothis; researchers claim that sprays can be reduced from seven to one with the same yield achieved (CC 1996, pp. 147–51; Mensah 1996).

If Bt cotton is to be a platform for IPM, something will have to give. To embrace food sprays, for example, farmers would have to move away from monoculture farming and broad spectrum sprays which disrupt the biological diversity required by the system. In short, it will take more than a transgenic crop to move farmers towards a broader-based IPM. Social factors will be just as important in shaping the future of this technology.

THE SOCIAL CONTEXT OF BT TRANSGENICS

A 'pesticide driven mindset' (CC 1996, p. 96), encouraged by the relatively cheap cost of sprays, may be one of the biggest threats to the Australian cotton industry. It may also undermine those producers seeking to introduce a truly diverse range of pest management tools. Conventional cotton farming, some argue, needs a paradigm shift. 'Attitudes must change! The idea that "the only good bug is a dead bug" must give way to an appreciation of the ecological role of a pest species and acceptance of their presence' (CC 1996, p. 153).

A familiar sentiment has been echoed by the late Elaine Brough, a Queensland-based entomologist who has advocated that producers overcome the risk of crises in pest management by moving away from a 'control focus' to a 'management perspective' (Brough *et al.* 1995). Her concern was that 'the promotion of new biotechnological tools as environmentally safe and effective threatens to displace IPM programs and cause a return to reliance on a single pest management method'. Whilst suggesting that biotechnology had a lot to offer pest management, she argued that we might be 'stepping off the chemical pesticide treadmill onto the biotechnological treadmill' (Brough *et al.* 1995). Colleague Myron Zalucki also believes Bt cotton is seen as a 'silver bullet' to substitute failed chemical control agents. 'The current view of IPM is sample, spray and pray it works until the next tech fix comes along' (Zalucki 1997). This mindset has been encouraged by the language of Monsanto sales documents for Bt cotton which have offered farmers 'peace of mind knowing that crops are continuously protected' (Monsanto 1995).

Fundamentally, the biological component of IPM requires the careful understanding and manipulation of complex ecological interactions. It is an approach that so far has not tended to fit easily with the agribusiness imperative to develop patentable technological solutions. Rhône Poulenc's food spray product is one thing, the people skills involved in making the whole system work is another. Says Zalucki of IPM, 'It's as much people management as pest management', and he doubts we currently have the management systems in place to cope with such a sophisticated approach (see Zalucki 1997).

In the 1996–97 season, for example, Bt cotton itself required 50 per cent more time to check than conventional crops and agricultural consultants could hardly meet the demand (CRDC 1997; *Australian Cotton Outlook* 1997, p.1). Some may well wonder if the industry is suitably equipped to cope with a labour-intensive biological control system based on food sprays, especially for transgenic Bt crops where large-scale regional co-ordination is essential (Zalucki 1997).

The fact that Bt has now been engineered into around 40 other crops awaiting commercial release is an important factor here (McLean *et al.* 1997, p. 243). The introduction of Bt genes into crops makes up 95 per cent of the commercial research into biological insecticides (Busch *et al.* 1991). Apart from cotton, Bt has been engineered into most of the major food and cash crops such as tobacco, corn, apple, canola, potato, rice and tomato (*Gene Exchange* 1992). While cotton is currently the only commercial transgenic Bt crop in Australia, Bt potato and corn have already been approved in the US.

Significantly, some crops such as corn and cotton share the same pests. This means that cornfields that might previously have provided populations of susceptible insects to dilute the resistant insects developing on cotton crops might, instead, provide more resistant insects! In other words, the growing numbers of Bt crops mean that we could wear out our Bt genes even quicker than forecasted.[3] While some argue cotton is the most deserving of Bt genes (McLean *et al.* 1997, pp. 280–1), incredibly there has been little dialogue between the different industries that use Bt, little indication of leadership from regulatory authorities, and even less public consultation.[4]

CONCLUSION

As 'the age of chemicals blends into the age of biotechnology' (CC 1996, pp. 37–8), proponents of genetic engineering predict a bright future for agriculture. Despite the sales hype, however, we have very little assurance that one of the 'flagships' of agro-biotechnology—Bt cotton—will provide a long-term sustainable future for the cotton industry.

On the one hand there are the cotton producers, caught between the rhetoric of integrated pest management and the reality of crisis pest management. On the other, the agricultural biotechnology industry, caught between the problems of resistance, and the need to recoup their investment. While Monsanto says it would be prepared to withdraw Bt from sale to prevent resistance, history has shown that chemical firms such as Monsanto have been more than willing to provide a procession of 'disposable' technological-fixes. Given the realities of the marketplace, and our relative ignorance of ecological processes this could be Monsanto's preferred path.

Many farmers who have chosen transgenics may see little option but to continue on the treadmill of crisis pest management, but the impact of 'wearing out' Bt genes will go far beyond *their* farm gate. Farmers of edible food crops such as vegetables may end up using less benign products, and organic farmers will have lost one of the few insecticides they can use.

Who will benefit, and for how long, from Bt crops? While some may insist we are on the brink of another 'green revolution', will agriculture under the impact of genetic engineering really move to a more sustainable phase? If the last 'green revolution' is anything to go by, we may be at best simply leaving one agribusiness treadmill for another.

Athlete's at Sydney's 2000 'green' Olympics may well sport t-shirts made of Bt cotton but the true measure of the sustainability of transgenic insect-resistant crops will be judged in another arena.[5] The appropriation of Bt by the transgenic juggernaut now calls for every possible ounce of commitment to IPM by industry and government, with a strong co-ordinating hand by informed regulators to ensure a regional and cross-industry strategy for its use. This is a tall order, indeed, in the current era of free market economics.

NOTES

The research for this chapter took place between 1995 and 1997. The content was accurate at the time of writing in 1997.

1 In the US, 40 per cent of growers who planted the first commercial crop of Bt cotton had to spray to kill heliothis, contrary to what they had expected (Monsanto n.d.). While the dose of toxin in Bt cotton was sufficient to control two US pests, it was insufficient to provide such effective control for another—*H. zea* (Prakash 1997; MacLean *et al.* 1997, pp. 259–72) (which, like the Australian heliothis pests, was naturally more tolerant to Bt). Some farmers suffered losses due to infestation from *H. zea* (Hutton 1997; MacLean *et al.* 1997, pp. 259–72). Fears are now held for

the longevity of Bt's effectiveness against *H. zea* (Roush 1997a; Prakash 1997).

2 To change this dose would require re-engineering Bt cotton. While CSIRO molecular biologist, Danny Llewellyn has some 'hunches' about how to fix the problem, he accepts that 'it's a big business world' and he cannot just experiment with the Bt gene unless the owner (Monsanto) allows him to (Llewellyn 1997). The company may well want to recover costs from the first version before lashing out to solve a problem in a relatively insignificant market such as Australia (Roush 1997a).

3 In the US, acres of corn which previously provided a refuge for *H. zea* are now being planted out with Bt corn from one of Monsanto's competitors (Roush 1997a).

4 Neil Forrester (1997) said that he tried to co-ordinate cross-industry resistance management with pyrethroids fifteen years ago—and gave up five years later. He said it was too difficult to do this—it was like 'taking the lolly off the kid' after the fact. He sees the only hope for transgenic Bt crops is for regulatory authorities to get in early at the registration stage. Despite calls from researchers (for example, MacLean *et al.* 1997, pp. 273–84), there are no visible initiatives from regulators on this front.

5 Sydney's successful bid for the 2000 Olympics was 'sold' largely on environmental grounds.

13

Throwing up concerns about novel foods

JANET NORTON

It was in 1953 that scientists Watson and Crick first described the double helix of DNA. This discovery, along with Mendelian genetics, paved the way for modern microbiology, biochemistry and molecular biology. From these disciplines—especially the latter—has come genetic engineering, and with it the capacity for scientists to move genes between species. In the words of one writer: 'genetic engineering enables us in principle to arrange the marriage and combine the qualities of any two creatures—or indeed of any number of creatures—into one, irrespective of species or indeed of kingdom' (Tudge 1993, p. 206).

With genetic technology, genes may be removed from any organism and moved 'across' species. Genes from a plant can be placed into an animal, or can be moved from an animal to a plant, or bacteria into plants or animals, or from plants and animals into bacteria. All these permutations—and more—are now possible. Genetic engineering allows scientists to choose the trait that they wish to express and, utilising the sophisticated tools of the modern laboratory, transfer it to another organism. Unlike the critics, and a growing number of dissenting scientists (for example, David Suzuki—see Introduction), many genetic engineers argue that recombinant-DNA technology is simply an extension of traditional breeding methods, albeit one providing greater 'precision' in genetic transfer and allowing new organisms to be developed in a shorter space of time. Scientists have been working on genetic engineering for decades. Now the results of their work are being commercialised, especially in food production.

Within Australia, cheeses are being made from biologically-engineered starter cultures, food additives such as flavouring agents and sweeteners have been genetically altered and products are available that have been derived from animals that have been treated with genetically-modified hormones and vaccines. Other products are also being developed and trialed. Transgenic tomato plants have been developed to have a longer shelf life; potatoes have been altered to prevent bruising and discolouring; pigs have been produced with enhanced growth due to the presence of a hormone derived from human sources.

In late 1996, Australia imported soy beans from the US which had been genetically engineered by the agribusiness company Monsanto to be resistant to the company's herbicide Roundup. Corn has been developed in the US containing the Bt gene (from a bacterium) which makes the plant insect resistant (see Salleh's discussion of Bt cotton in chapter 12). Plants are being developed to have a greater salt tolerance. Cattle are also being genetically modified so that they can more efficiently process the grasses that they eat. It has also been suggested that the genes of some vaccines may be introduced into fruit crops—providing those eating the fruits with immunity to specific diseases (Osfield 1997).

These are just a few of the hundreds of applications of genetic engineering that are currently being developed. These applications are moving out of the laboratory and are becoming available for consumption. The anti-ripening Flavr Savr tomato has been on the market in the United States for some time, but has now been withdrawn due to problems of bruising (*Gene Exchange* 1997). A Unilever version of it is currently being field-trialed in Australia. Roundup Ready soy beans have also been exported to Europe, as has Bt corn. It has been suggested that these new types of foods are necessary to maintain food supplies for the world's population. Further, bioscientists, economists and industry leaders have argued that these new genetically-engineered foods will increase Australia's competitiveness in the world marketplace.

The lack of immediate social acceptance of these new foods is sometimes dismissed by Australian scientists and representatives of government agencies. They believe that any consumer resistance to these products is the result of ignorance. They argue that without sufficient knowledge of the biology of genetic engineering consumers may, in the short term, be unduly suspicious of new scientific advances. According to some prominent geneticists, suitable education is viewed as the mechanism to dispel the public's supposedly 'irrational' fears (Peacock 1994; as discussed also in chapters 3, 10 and 11). In contrast, social research is increasingly showing there is growing public disquiet towards biotechnology.

RESEARCH INTO CONSUMERS' ATTITUDES

To investigate consumers' attitudes to genetic engineering, a number of studies, employing a variety of different methods, have been conducted overseas. In an effort to provide both in-depth and cross-sectional information about consumer preferences, a variety of research techniques such as surveys, focus groups and consensus conferences have been used. For example, in the US the Office of Technology Assessment (OTA) carried out a telephone survey in 1987. The survey revealed that:

> . . . while a majority of the public expresses concern about genetic engineering in the abstract, it approves nearly every specific environmental or therapeutic application. However, while Americans find the end products of biotechnology attractive, they are sufficiently concerned about potential risks that a majority believe strict regulation is essential. (OTA 1987, p. 5)

Since 1987, a number of other US studies have been conducted. They indicate that there is, indeed, consumer resistance to genetic engineering. A North Carolina study showed that many people felt left out of the decision-making process when new technologies were introduced (Hoban 1989), and that people showed more concern about the manipulation of animals than plants. In 1993, a telephone survey in New Jersey produced results similar to those from North Carolina (Hallman and Metcalfe 1993). In this research, participants regarded scientists as the most credible source of information on genetic engineering, while companies involved in genetic engineering were rated as the least credible. Focus group research by Zimmerman *et al.* (1994) established that, in a number of States of the US, the public demanded reassurance that genetically-engineered foods were safe, as well as information about the way in which they were being produced. This study again showed that participants had only moderate trust in statements on biotechnology made by government agencies and were not totally convinced that the government could protect them from the hazards of these new foods. These studies also indicated that some participants found the new technology unacceptable because of 'moral beliefs' (Zimmerman *et al.* 1994).

Canadian research produced similar results. Participants considered gene transfer to be acceptable only if there was a specific goal to be achieved. As well, attitudes to genetic engineering were found to be affected by a respondent's 'core beliefs'—that is, their knowledge of science and technology, attitude to nature, and religious beliefs. The study concluded that respondents had no clear understanding of biotechnology and most felt unable to make absolute judgements about the technology. This research reported that the key strategic imperative of biotechnologies' proponents will be to convincingly address the concerns and allay the fears of Canadians around the issue of 'social control that was . . . described in the study' (Decima Research 1993, p. 34).

Within the European Community several sequential surveys have been conducted into public perceptions of genetic engineering. The so-called 'Eurobarometer' is conducted every few years and examines attitudes to a number of genetic-engineering applications (INRA 1991). Since 1991, support for genetic engineering has been declining (Jank 1995). This is despite the fact that comparisons of the 1996 Eurobarometer with the 1993 and 1991 surveys show that the public's knowledge has increased. Respondents now appear to be less optimistic that applications of genetic engineering will improve their quality of life. This study also indicates that respondents are concerned about the potential risks of genetic engineering and the 'moral acceptability' of some applications (Wagner *et al.* 1997). Respondents also considered that genetically-engineered foods should be labelled, and that there should be public consultation prior to the development of new biotechnologies.

In another study—conducted in the United Kingdom—biotechnology was perceived to have low benefits, high risk and low need. Respondents exhibited a lack of confidence in business and government organisations to regulate genetic engineering. The type of genetic engineering (for example, whether it was conducted with animals or bacteria) determined the amount of support that it received. Genetic engineering of microorganisms and plants was more acceptable than the genetic engineering of farm animals (Sparks *et al.* 1994). In other British and European research, lack of information on genetic engineering, environmental considerations, and the lack of public participation in decision making, regulation and control were discussed by both proponents and critics of genetic engineering. Researchers found that it was almost impossible to have dialogue between these two groups. This study concluded that one of the main issues arising from the current debate was the need for future research to be undertaken to determine attitudes and preferences of the public, specific concerns of the public, and the relationship between the understanding of science and perceptions of biotechnology (Lemkow 1993).

Closer to home, at the beginning of this decade the New Zealand Department of Scientific and Industrial Research conducted research into public attitudes to genetic engineering. The 1990 survey found that while most respondents were familiar with the term 'genetic engineering', they could not adequately explain what it involved. Concerns were expressed about the potential for disaster, possible hazards arising from unknown areas of research, problems of control, environmental damage, and the unknown side effects of genetically-engineered foods. Manipulation of plant cells was considered to be most acceptable—manipulation of human cells the least acceptable. Concerns identified were centred upon the 'moral and ethical implications of the research and on the possibility of environmental or human hazards resulting from the introduction of genetically modified organisms' (see Couchman and Fink-Jensen 1990, p. 7).

The findings from these studies have several commonalities. There is a lack of real understanding of what genetic engineering is; genetic engineering of microorganisms and plants is viewed as being more 'morally acceptable' than the genetic modification of animals; and respondents wish to have genetically-engineered products labelled. A deeper look at the underlying factors affecting attitudes to genetically-engineered products also shows that, in general, women, older people and those with lower educational levels, are less likely than others to favour genetic engineering.

In Australia, there has been little research into the public's attitudes to genetic engineering. The studies which have been published tend to indicate that Australian consumers have some—but not all—of the same concerns as their northern hemisphere counterparts. For example, following an exhibition produced by the CSIRO on genetic engineering, Alexander (1992a) analysed the public's response. The majority of those who answered questions after seeing the exhibition indicated that they had major concerns in relation to scientists interfering with nature. They were concerned that control might be exerted by companies commercialising the new products, and that there might be risks to the environment. This was despite the fact that the exhibition was largely pro-biotechnology (Alexander 1992a; Hindmarsh 1992).

But, in another study, commissioned by the Department of Industry, Science and Technology (DIST), and conducted in 1994, Australians demonstrated some support for genetic engineering. A number of applications—including new drugs, pest-resistant cotton, cooking oils, lean pork and a tastier tomato—were described in the survey (see Kelley 1995). Contrary to overseas studies, there was a positive reaction to genetically-engineered products. The study also showed that there was no variation in support in relation to gender, age or educational levels. Opposition to genetically-engineered products was exhibited only by those who placed a low priority on improvements in health and agriculture and those without a scientific world view. However, results did indicate that respondents wished to see any new products labelled (Kelley 1995).

Given that these results appear to be in sharp contrast with overseas findings, it is important to note that the survey instrument that was employed in this research has been criticised for having provided somewhat superficial and biased information to the respondents (see Hindmarsh *et al.* 1995). For example, in one statement to respondents it was implied that opposition to genetic engineering was the province of 'greenies' or radicals. Particular products were presented to the public in a positive light. Thus the survey reported the prospect of 'healthier' cooking oils, 'leaner' pork and a 'tastier' tomato. No indication of the potential risks of these products to either the environment or personal health was made within the survey. The survey implied that all products would be labelled as genetically modified and it is on this assumption that respondents replied (see Hindmarsh *et al.* 1995). Yet, labelling of these products is not presently undertaken in Australia and there is resistance on the part of companies to do so.

Despite the reported support for genetic engineering, other results from the DIST survey demonstrated that respondents did have concerns about genetically-engineered foods. Ninety four per cent of respondents indicated that they were worried that genetically-engineered plants might spread on their own. A similar percentage were concerned about the long-term health effects of eating genetically-engineered foods. With these concerns in mind, the pro-biotechnology results of the 1994 survey must be seriously questioned.

FURTHER RESEARCH IN AUSTRALIA

In 1995, a series of focus groups were conducted in Rockhampton, Queensland. The aim was to obtain some understanding of consumers' knowledge of—and attitudes to—genetically-engineered foods. Participants talked freely about what they considered important when choosing and buying food. The topic of genetic engineering was then introduced by providing participants with information on a number of genetically-modified products—both foods and non-foods. This information not only identified the product but the *type* of genetic manipulation that was used to produce it. Participants were given the chance to discuss this information and to ask questions. This methodology demonstrated that participants had an unclear picture of what genetic engineering was and how it was being used to produce new foods. In the words of one participant:

> What is genetic engineering and what does it actually mean, what are the consequences, what is an enzyme or something else that's in there, you know? Like . . . people don' t know.

Participants were asked to indicate their degree of support for the various products which had been described. In the course of discussions, a number of concerns about genetically-engineered products were expressed, the primary one being concern about the genetic alteration of foods. Participants were concerned about the short- and long-term health effects that ingestion of these foods might produce. They were particularly concerned for future generations and felt that they would be the ones who were most at risk from problems caused by this technology:

> But how do you know it's not going to mutate and then in the future we might mutate? It might sound stupid but we don't know what kind of effect these are going to have down the track.

As well, participants wanted more information about the nature of genetically-engineered foods. They certainly demanded that all genetically-engineered products be labelled. Genetic engineering of animals to improve their quality of life received qualified approval as long as products from these animals did not reach the dinner table (for example, meat from the blowfly-resistant sheep). As in previous studies, genetic engineering of plants and microorganisms was perceived as being preferable to the genetic engineering of animals (Norton and Lawrence 1996).

Based upon the information gathered in the focus groups, a survey was developed and administered nationally during the period December 1996 to January 1997. The mailed survey—called here for convenience the Central Queensland University (CQU) survey—provided respondents with information on genetic engineering in general as well as a number of specific products. (Products described were: a tomato developed using other plant genes; cheese produced using genetically-engineered rennet; wheat made pest resistant by insertion of genes from a bacterium; a blue rose developed using the insertion of genes from another plant; leaner pork developed using the insertion of genes of human origin; sheep which acquired resistance to blowfly strike through the insertion of plant genes; and a tomato developed by inserting fish genes.) A number of questions relating to respondents' knowledge of science and acceptance of science and technology were also included.

The results are now becoming available from this survey. They show—in line with overseas surveys—that respondents found it more acceptable to genetically-engineer plants (68 per cent) than animals (40 per cent). The use of human genes in other organisms and the genetic engineering of humans proved to be unacceptable to over 70 per cent of the respondents. There was clear support for government control of the technology and for the labelling of genetically-engineered products. A majority of respondents (53 per cent) considered that, in general, the risks of genetic engineering outweigh the benefits. But, when looking at individual products, in all cases except the example of leaner pork using a human gene sequence (where support dropped to below 35 per cent), a small majority considered that the benefits outweighed the risks. While 22 per cent of respondents considered that genetic engineering poses no risk to society, a larger percentage—some 28 per cent—considered that it offers no benefits at all. Respondents also had some concerns about environmental damage that might be caused by release of these new organisms and a majority considered that eating these foods could cause long-term health effects. In the words of a typical respondent:

> I do not believe that benefits will arise from humans manipulating genes of plants, animals or humans—particularly the last two—because it is not known what long-term effects will be for the human race who eat a lot of genetically-altered foodstuffs.

The risk to the environment of accidental release was also a concern—with over 70 per cent of respondents considering that the accidental release of genetically-engineered organisms will cause environmental damage.

> Science has a place in modern society but not at the cost of the environment. The environment and the long-term effects of science on the environment should be priority number one.

Respondents were also asked to indicate their support for the development of the various products listed above—if those products were to be produced using *conventional* means (that is without the use of genetic engineering). In all cases a majority of respondents supported such development, with the greatest support being for the development of a variety of wheat that requires reduced amounts of insecticide spraying (95 per cent) and the development of blowfly-resistant sheep (93 per cent). The support for all other products was well over 70 per cent. However, support for the development of all products *decreased significantly* when respondents were asked to consider the development of these products using only genetic engineering methods. With the exception of the two previously-mentioned products, support was recorded at less than 40 per cent. It can therefore be argued that concern over these products is related to the *type* of technology used to produce them and *not* related to the product itself. A typical open-ended response was: 'It is an ongoing experiment, the end results of which cannot be predicted or controlled'.

In all cases that concerned food, at least 35 per cent of respondents indicated that they would not buy the product. This rose to a high of 65 per cent of respondents who indicated that they would not buy genetically-engineered pork. However, with the exception of the cheese produced with genetically-engineered rennet (55 per cent do not appear worried about this), a majority of respondents indicate that they had concerns about eating the product (55 per cent worry about eating the wheat and tomatoes, and 64 per cent worry about eating the genetically-modified pork). For one respondent:

> All my life scientists have been saying this will not harm humans only insects, rabbits etc. Ten, twenty years down the track we find it is

> accumulative and does quite a bit of harm. So now we invent something to fix that and cause another problem. So I say, enough is enough. Please leave something in its natural state for us to enjoy with a bit of quality as well as taste.

Survey respondents also exhibited a lack of knowledge of the type of research being conducted, with over 50 per cent indicating that—up until the time of the survey—they had heard or read little or nothing about genetic engineering methods. This rose to over 60 per cent of respondents who indicated that they have little or no knowledge of the particular types of research that are being undertaken. However, two-thirds of respondents felt that their knowledge of science is adequate or better than average (70 per cent had received science education at secondary school level) and that they read about science or watch science shows on television with some regularity. In open-ended questions, respondents repeatedly reported their lack of knowledge of genetic engineering and its effects (a point analysed politically in chapter 3), as well as the lack of information available to them:

> The problem to me is there's not enough information known either way (about it) being safe or good. I think a lot more study and testing needs to be done to make it safer for consumption.

Further analysis of the data showed that women and those believing in God found genetic engineering of plants and animals less acceptable than do men and those who do not believe in God. There were no differences among demographic groups for acceptability of the genetic engineering of human beings: all groups being opposed to it. Women, and those with a belief in God, were also less likely to perceive any benefits in genetic engineering.

THE RISK SOCIETY

Society in the late twentieth century has experienced the unexpected—and sometimes deleterious—results of many technologies. The air pollution caused by factories has had to be controlled by new emission laws. Nuclear technology has resulted in the Three Mile Island and the Chernobyl accidents. Hailed as the saviour of agriculture when it was first introduced, DDT has been shown to cause problems for animals high in the food chain, and its long-lasting residues have caused on-going problems for farmers. Lead in paint and car exhausts has had to be removed to protect our children. The increased use of electrical energy produced by burning fossil fuels has helped to increase the greenhouse effect. The installation of sewage systems in our cities and towns has left us with waste disposal problems. Food additives once deemed safe have caused allergic reactions in many people and have resulted in the stringent labelling of food.

All of these technologies were developed to meet perceived needs in society. Whether these needs were largely economic, or social, or political is open to debate. And, while it is obvious that technologies will impact in unexpected ways, some have produced 'side-effects' which have caused major environmental and health problems. In other words, the side-effects are no longer simply a threat to our income and employment as they were seen to have been in the early nineteenth century. They now threaten social well-being. Thus, for one CQU survey respondent:

> Science can and will progress rapidly, but the results of such research need to be carefully reviewed and regulated, as complex technology has the inherent capacity to cause complex problems also.

In his influential book *Risk Society: Towards a New Modernity*, Ulrich Beck (1992) argues that the problems of technology are recognised by society, not on a small scale, but in terms of the potential for large-scale devastation and destruction. Devastation on this scale can only be a threat, not a reality; for it to become reality would mean total annihilation. This does not mean that it is viewed any less significantly. Indeed, that the risks are largely unknown makes people all the more concerned. While the public recognises the risks, at the same time, people experience a lack of control over the assessment of the risks. The task of assessment is placed in the hands of government and science—often the ones who have introduced, or endorsed, the technology in the first place.

Beck argues that invisible hazards are becoming more visible, adding to the cumulative effects of all hazards. Forests are seen to be disappearing, toxic accidents are becoming more common, and our foods have increasing levels of additives. As well, these events are becoming increasingly visible through the media. Not only is actual risk increasing—so is the perception of risk. Scientists argue that this perception of risk is the result of ignorance or a lack of knowledge, and that, *ipso facto*, all that is required to overcome public resistance is education. Beck argues that the reverse is true. As the public becomes more knowledgeable, the greater its perception of risk. He further argues that the current climate of criticism of science and technology is linked to public cynicism—and often rejection—of techno-scientific rationality. This coincides with a growing demystification of science.

Respondents to the Australia-wide CQU survey demonstrated this perception of risk as well as a distrust of modern science. In the written words of one respondent:

> Whilst I trust scientific research, I do not believe anybody can give complete assurances of the safety or reliability of anything to do with genetic engineering. There are too many instances of the unexpected happening in science. You have only to look at the side-effects of supposedly safe drugs and vaccines which have killed and crippled many people. Then there is the cane toad, introduced on scientific advice (although well-intentioned); the present concerns about plastics on male fertility; the current theory about the origin of the AIDS virus in humans (from polio vaccine). We cannot foretell all consequences of technological advances.

CONSUMER RESISTANCE TO GENETICALLY-ENGINEERED FOODS

Research has demonstrated that consumers, both overseas and here in Australia, have concerns about genetic engineering. While research has shown that there is less concern about the genetic modification of plants than there is about the genetic engineering of animals, few studies have actually asked participants if they will eat the new foods. The CQU Australian survey indicates that respondents do have concerns about eating genetically-engineered foods—both plant and animal. Over half of respondents are worried about eating these foods and the long-term health effects resulting from their ingestion. Respondents also consider that products should be labelled as being genetically modified.

While the products of this technology are only now becoming commercially available within Australia, they have been available for some time in the European market, where there has been resistance to the new food products. For example, the importation of genetically engineered corn and soy beans into Europe resulted in opposition from many sources. Greenpeace mounted blockades and there was a public outcry about the new products. While the European Commission initially proposed that genetically-modified grain be segregated from conventional grain it will now permit mixed shipments with the requirement that those containing genetically-engineered grains be labelled as such. In Japan, Kagome Company has suspended marketing processed foods from genetically-engineered tomatoes. This is a direct result of pressure from consumer groups (Purefood 1997). Japan's public opposition to genetically-engineered foods also has resulted in Canada and the United States facing reductions in their market share because they are known to grow transgenic crops. Instead Japan is looking to Australia to obtain conventional grains (Bolt 1997; McNulty 1997).

EuropaBio, which represents the interests of the European biotechnology industry, is becoming so concerned about the effects of public pressure that it has engaged crisis management consultants, Burston

Marsteller (the company which represented Babcock and Wilcox after the Three Mile Island incident), to improve their image (as proponents in Australia are also actively attempting to do—see chapter 3). Burston Marsteller is working to change the public's perceptions of genetic engineering. It has advised the industry to focus on 'symbols, not logic' in overcoming resistance, as it considers the industry will be unable to win the arguments over the risks posed by genetically modified food—particularly those relating to environmental dangers (see Penman 1997). Research into European attitudes to genetic engineering showed similar results to those obtained in the CQU survey. Australia can therefore expect to see similar resistance as these products enter the marketplace.

CONCLUSION

Genetically-engineered foods are gradually entering the world marketplace. Already genetically-engineered grains and other foodstuffs are becoming available overseas and Australia has received its first shipment of genetically-engineered soy beans. Scientists and industry are working on several new ways of using this technology. When this technology is applied to foods and food products, it will result in products of uncertain genetic pedigree. Without labelling, consumers will be unaware of the role of genetic engineering in producing these food products. Yet, consumers are becoming increasingly concerned about the quality of the food they eat. Furthermore, there is a growing awareness of the risks that modern science and technology pose to people, to the environment, and to future generations. Science is no longer accepted unquestioningly as providing benefits to society. Research, both in Australia and overseas, indicates that consumers are becoming more aware of the risks that genetic engineering poses. These risks are seen to have the potential for environmental damage caused by the release of genetically-engineered plants and animals—as well as the possibility of long-term health effects caused by eating genetically-engineered foods. Consumer lobby groups are asking for government control of the industry and labelling of genetically-engineered products.

Consumer resistance is not directed at the new food products intrinsically, but rather at the *type* of technology used to produce them. If conventional means are used, the products are acceptable to consumers. Acceptance drops, however, if they are to be produced using genetic engineering. Scientists have argued that lack of knowledge of genetic engineering is the cause of resistance to the technology. However, even when people are provided with examples—and with explanations which enhance their knowledge—support is not forthcoming. Research has shown that while the genetic manipulation of plants and microorganisms is more acceptable than for animals, consumers still express a reluctance to eat these products. There is, however, support for some applications of genetic engineering—in cases where the resultant product is not to be eaten, and particularly where it enhances animal welfare. Consumers are concerned about the lack of information on genetic engineering and the potential for environmental damage and long-term health problems. Consumers want government control and labelling. In addition, some countries are looking to Australia to provide transgenic-free crops. Accordingly, it may be prudent for Australia to institute regulations to protect consumers and to ensure that all genetically-engineered foods are properly labelled.

14

Genetic engineering: the campaign frontier

BOB PHELPS

Brisbane, Australia, 1 December 1996. In the dead of night a bulk carrier sidles quietly up to the remote grains terminal on Pinkenba wharf near the mouth of the Brisbane River. In the quarantine area behind a stout chain link fence a controversial cargo is secretly unloaded under strict security. It must remain isolated from the Australian environment. The security perimeter also excludes numerous protesters, angry over potential threats to the environment—and to food quality and safety—posed by the shipment. They have been calling for weeks on both federal and state governments to ban the shipment and its load but Ministers have refused to intercede. The shipment is transferred to the local factory of American seed giant Cargill for processing into human and animal food products. Where or when the food is eaten no-one knows as Australia has no enforceable rules on the assessment, monitoring or labelling of novel foods (Ripe 1996).

The protesters are members of both the Australian GeneEthics Network and other public-interest groups, as well as concerned individuals. The Network—sponsored by the Australian Conservation Foundation—is a federation of groups and individuals promoting critical debate and education in relation to the impacts of genetic engineering. The contentious cargo contains Monsanto's 'Roundup Ready' soybeans, the first genetically-engineered whole food to enter Australia's food supply. Monsanto used gene technology to make a designer soy crop that tolerates the company's own top selling chemical—glyphosate—with the brand-name Roundup, a broad spectrum herbicide.

The risks and hazards of this strategy are borne by food buyers around the world. On medical advice and in response to advertising, soy has now become an integral part of Western diets. Soybeans, soy protein isolates, and other soy products from the US include Monsanto's 'gene-beans'. Soy is also in a majority of processed foods and other products such as vitamin pills and cosmetics, including vitamin E cream. But Monsanto's soy contains new substances never before in food—genes and proteins from other species—as well as increased synthetic chemical residues.

'Gene-beans' have no history of safe use, nor have they undergone pre-market human testing. At potential risk are the growing population of people with allergies, chemical sensitivities, immune suppression, lactose intolerance, and menopausal women, who have embraced soy products as a healthy staple. It seems clear from the gene-bean saga that companies and governments plan to deploy gene technology products without public consent.

GENETIC ENGINEERING INDUSTRIES

The pattern of gene technology innovations is to entrench chemical–industrial agricultural practices and extend corporate monopolies, by modifying organisms for intensive factory and farming systems. One aspect of globalised industrial agriculture—facilitated by genetic engineering—are foods designed to be traded over extended timeframes and to distant markets. Under this system, products must arrive at the point-of-sale looking presentable and fresh, long after their conventional counterparts would have decayed. They are intended to attract a premium price, to cover extra royalty, transport and handling costs, even if their nutritional and health value may be compromised. The genetically-engineered long-life tomato—the Flavr Savr tomato (also patented by Monsanto)—was one such attempt to design a vegetable which never appeared to age. Indeed, it was the first transgenic whole-food product available in shops (see Krimsky and Wrubel 1996, p. 100). Many of the tomatoes—introduced into the US market in 1994—were soft and bruised on arrival and could not be sold as fresh fruit (*Gene Exchange* 1997, p.12). This exemplified the issue of 'counterfeit freshness' in non-ripening crops.

Gene technology also offers industry a profitable replacement for many non-renewable resources which are becoming more scarce and expensive, especially the fossil fuels on which existing industries such as plastics, drugs and food depend. Corporate and government resources have been mobilised over the past fifteen years to modify—through r-DNA technology—the more abundant and potentially renewable stocks of plants, animals and microbes. They are viewed as the catalyst for expansion of many existing industries, as well as the new bioindustries. For example, animals have been engineered to express drugs and human proteins in their milk, plants can make plastic polymers, and a vanilla-bean essence can be brewed in a vat. The social and environmental impacts of such innovations are not currently assessed as regulations focus solely on narrow technical assessments (see chapter 3 for an analysis of how this has occurred).

A central promise of modern biotechnology is to feed the world, but this promise is like many other myths (for example, those of nuclear industries) created to assuage public concern and encourage acceptance of dangerous and inappropriate technologies. The benefits have been strongly promoted, while the costs, hazards and risks have been played down or ignored, despite abundant warnings of insidious effects well before the technologies were deployed. Many promised pay-offs of mega-technologies have never materialised or have been short-lived, and many impacts are still not officially acknowledged (see Hynes 1989; Beck 1995). This and future generations have already been saddled with radioactive wastes and toxic chemicals, in addition to a host of other adverse impacts from the widescale introduction of ecologically-unsound and socially-unjust technologies. A similar pattern of reckless promotion and use is now being encouraged through manipulative promotional campaigns extolling the merits of a bioengineered world.

LIFE PATENTS AND MONOPOLY CONTROL

Monsanto's gene-beans are a patented product, protected under US Patent numbers 4 635 080, 4 840 835 and 5 352 605. Under contracts on the Roundup Ready seed/chemical package, farmers are required to use only Monsanto's Roundup; are prohibited from saving seed for future planting; are responsible for any breach of the agreement for up to three years after harvest; and the company is allowed to enter farms without notice to remove plants if it suspects there has been a breach (Monsanto 1995). Such powers reflect increasing monopoly control over food production by corporations, facilitated greatly by *Plant Breeders Rights and Patent Acts*.

In Australia, patents for life forms were permissible as early as 1976 (Australian Official Journal of Patents 1976, p. 3915). This was reiterated in 1980 by the Patent Office stating that: 'no distinction is to be made solely on the basis that a claimed product or process is, or contains or uses, a living organism. Higher life forms will not be treated any differently from lower life forms such as microorganisms' (cited in Lee 1992a, p. 225). The range of patentable subject matter in this area includes: microorganisms, cell lines, hybridomas and other related biological materials including bacteria, fungi, algae, viruses, fermentation or enzyme using processes to synthesise a desired chemical compound or composition, mutation or genetic engineering, and the use of microorganisms to produce food or beverages. Patent protection can also be obtained for 'inventions' including, among other things, genetically-modified bacteria, plants and non-human organisms; synthetic genes or DNA sequences; mutant forms and fragments of gene sequences; the DNA coding sequence for a gene; the protein expressed by the gene; host cells carrying the gene; higher plants and animals carrying the gene; and perhaps, most significantly, the building blocks of living matter such as DNA and genes (including human DNA and genes) which have for the first time been identified and copied from their natural source, then manufactured synthetically as unique materials with a definite industrial use. DNA or genes in the human body are not patentable as such, however, a gene sequence which has been separated from the human body and manufactured synthetically for reintroduction into the human body for therapeutic purposes would be patentable. The treated human body is not patentable (Australian Industrial Property Organisation 1997, p. 1).

In December 1994, scientists from the Howard Florey Institute at the University of Melbourne were at the centre of the world's first legal challenge against the patenting of human genes. The contested patent application was for a gene that codes for human relaxin, a hormone primarily produced by the ovary during pregnancy which softens the cervix and reshapes the birth canal for delivery. The Institute's scientists also found that relaxin has powerful effects on heart tissue. A genetically engineered form of the hormone might ease labour or treat fetal cardiac problems. The bio-products—if successfully developed—would be worth millions of dollars (Ewing 1994, p. 3). The patent application raised a challenge from the Greens in the European Parliament about its legality, on the grounds that human genes should not be transferred into 'private ownership' and that any commercial use of the gene would be immoral:

> The publication of this alleged invention arouses revulsion at the prospect of the genetic material of human beings—the common heritage of all humankind (or in special cases of certain individuals)—being carved up into its components and, through an administrative act of the European Patent Office, transferred to the patent holder for its exclusive economic exploitation. A demonstration that such an act offends moral principles may be found in the religious, tradition, and/or ethically-based indignation of every person who is or will be actually informed about it—excepting those who themselves belong to the gene-industrial complex (Greens in the European Parliament n.d., p. 3).

Dr Lynette Dumble, a senior researcher at the University of Melbourne's department of surgery also took a different position from her colleagues at the Howard Florey Institute. She said that the gene 'belongs to humanity' and that granting the patent would lead the use of the hormone for treating disease to become commercially restricted (Ewing 1994, p. 3). The Institute was granted international patents for the process of synthesising human relaxin and its fusion products in June and July 1994 (University of Melbourne 1994).

Many other Australians also oppose life patents including the GeneEthics Network and the Australian Conservation Foundation which argue: 'Patents on transgenic organisms (animals, plants, microorganisms or human beings), their genetic material and processes for creating them, should not be permitted' (ACF 1989). It is our view that living organisms are not inventions and should not be privately nor monopoly owned. In June 1996, the Australian Democrats introduced a Private Members Bill into the Senate to amend the *Commonwealth Patents Act* 1990 to prevent it from applying to 'naturally occurring genes', 'naturally occurring gene sequences' and 'descriptions of the base sequence of a naturally occurring gene or a naturally occurring gene sequence'. In mid-November 1997, with the proposed Bill still awaiting Senate discussion, Democrat Senator Natasha Stott Despoja set out her ethical view in the *Canberra Times* (Stott Despoja 1997, p. 43) that 'life forms cannot be owned like other objects of commerce'. Sharing the sentiments of the contributors to this book, she went on to conclude 'the Parliament should also encourage a broad-based debate over gene-technologies' (Stott Despoja 1997, p. 44).

Such opposition to life patents is met with extreme resistance by the gene-industrial complex as the global patents system is the main institutional mechanism used by Northern-based companies to reap the commercial benefits of gene technology, with royalties constantly flowing back to their head offices. For example, the seeds of Monsanto's Bt cotton being grown in Australia (see chapter 12 for a thorough assessment of Bt cotton), bears a technology licence fee of some $245 per hectare. The US government pressures countries which resist biological patents, saying refusal is a non-tariff barrier to free trade, and subject to possible sanctions under World Trade Organisation rules (see also chapter 4).

During 1997, the Clinton Administration intervened in Thai Government plans to protect indigenous resources, skills, culture and local production. A letter from the US Department of State to the Royal Thai Government dated 21 April 1997 said of draft laws allowing Thai healers to register traditional medicines: 'Washington believes that such a registration system could constitute a possible violation of the TRIPS (Trade Related Intellectual Property Rights) and hamper medical research into these compounds' (cited in Dawkins 1997). That Thailand is not obliged to comply with TRIPS until at least the year 2000, and that medical practices may in any case be exempt, apparently had no influence on the super-power. A global coalition of over 200 concerned groups called on US Secretary of State, Madeleine Albright, not to interfere with the legitimate protection by other nations of their cultural and biological diversity (see Dawkins 1997). To protect and expand sustainable systems, traditional farmers and conservationists need to maintain agricultural diversity and low-impact farming practices. In strengthening corporate control of production systems through patents, a parallel effect is to marginalise and extinguish traditional, indigenous or local knowledge and practices. Global industry and Northern governments have redefined and misappropriated the existing patent system to legitimise biopiracy (see also chapter 4), so they can gain monopoly control of the whole living world.

REGULATING GENETICALLY-ENGINEERED PRODUCTS

The genetic engineering industry and governments are fast-tracking genetically-engineered products into the environment and marketplaces—with minimal regulation (the latter aspect is one that Hindmarsh analyses in chapter 3). Monsanto and Australian regulators secretly discussed Roundup Ready soybeans for months before the then Science and Technology Minister, Peter McGauran, disclosed on the ABC's television programme *Landline* in October 1996, their imminent arrival. Federal authorities permitted Cargill Oilseeds to import the beans without the usual public notice or consultation, without formal assessment, and without a genetically-engineered food standard in place.

In early July 1996, Monsanto had asked the Genetic Manipulation Advisory Committee (GMAC) to rule on Roundup Ready soybeans, the first step in getting import approval. The Committee normally gazettes all proposals and allows 30 days for public comment but did not do so. It also ignored the standard practice of requiring field trials in Australia to demonstrate safety and efficacy prior to general commercial release. To make matters worse, the favourable GMAC decision appeared to be based only on data generated overseas by Monsanto itself.

Quizzed later, GMAC's secretariat said public notice was unnecessary as quarantine rules require all imported soybeans to be taken directly from wharf to factory for processing, excluding them from the Australian environment. Only after public controversy had surfaced over the first shipment, did GMAC say publicly that the beans posed no genetic hazard and therefore should be allowed. GMAC's advice went to the Australia New Zealand Food Authority (ANZFA), which also ignored its usual practice of notifying the public. Because there was no genetically-engineered Food Standard in place, ANZFA said that it was beyond its power to assess, approve, or register the beans.

Before the first gene-bean shipment landed in Brisbane, the GeneEthics Network expressed its reservations to the Minister for Primary Industries and Energy, John Anderson, and asked for a ban. Failing that, the Network wanted the Quarantine Inspection Service, under the powers of the Imported Foods Inspection Program, to test all shipments for illegally-high Roundup residues. This was essential because of the gene-beans and the new US practice of defoliating conventional soy crops with Roundup just prior to harvesting. Monsanto applied to the National Registration Authority on Agricultural and Veterinary Chemicals (NRA) (the federal body responsible for all chemical registration and standards of use) to increase residues in imported soybeans from 0.1 mg/kg to 20 mg/kg. This suggested that tests—if carried out—would have found Roundup over the approved level of 0.1 mg/kg. The Minister refused to vary the standard sampling regime of one in twenty shipments for imported foods, so it may be a long time before residue levels are measured or illegal residues detected.

Like GMAC and ANZFA, the NRA did not call for submissions on Monsanto's application for a 200-fold rise in Roundup levels in imported soybeans. Defying logic, NRA officials claimed the proposed increase did not require public review because Roundup was already used in conventional soy crops. This ignored local and overseas prohibitions on spraying the chemical directly on crop plants, imposed because such use would kill them. Indeed, that is still illegal in Australia and the allowable residue limit remains effectively zero as the limit of 0.1 mg/kg is the threshold of detection. The NRA also said the proposed rise in glyphosate residues in imported soy was necessary to align Australia with the relevant food standard of the *Codex Alimentarius*—the UN body that sets global food production and processing standards—under World Trade Organisation obligations. At Monsanto's urging, the Codex standard had been earlier raised to 20 mg/kg—also without public knowledge or discussion.

Only in late November 1996—when the first shipment of gene-beans was about to arrive in Australia—did the NRA take the final step to legitimise the beans. The NRA recommended to ANZFA—on Monsanto's behalf—that the Food Standard be amended to allow the 20 mg/kg of glyphosate residues in imported soybeans. Just two lines in a news bulletin, and a government Gazette notice, told the public of the Monsanto/NRA proposal in January 1997.

Until a general standard on genetically-engineered foods is approved by all Australian and New Zealand Health Ministers, ANZFA decided not to consider the NRA request. Thus, potentially illegally-contaminated gene-bean shipments continue to enter Australia with Ministerial approval alone. Food regulation in Australia is certainly an *ad hoc* affair, and cannot be trusted to conform to established practices. Even token but standard processes of public consultation permitted under present rules were dispensed with, to facilitate the gene-bean entry into Australia.

STANDARDS REQUIRED FOR NOVEL FOODS

In early February 1997, ANZFA published Proposal P97—a draft standard on 'Food Derived from Gene Technology'. Largely due to much publicity about the gene-beans, ANZFA received over 3000 submissions from Australians and New Zealanders. In contrast, when the relatively unknown standard setting process began in 1993, less than 200 submissions were received. Most of the 1997 protest letters advocated a ban on the gene-beans (which endorsed the public's sentiments concerning r-DNA technology expressed in submissions to the 1990–92 parliamentary inquiry—see chapter 3). They also advocated complete labelling of all genetically-engineered foods (which confirms Norton's findings in chapter 13).

When assessing gene technology proposals, regulators must take a precautionary approach and assume that things can go wrong. The GeneEthics Network policy thus states: 'The onus should be on the proponents of genetic engineering applications to reveal all the potential hazards of their proposals and to show that the uses are benign, not on governments or citizens to demonstrate the dangers' (ACF 1989). Similarly, the parliamentary inquiry report—*Genetic Manipulation: the Threat or the Glory?*—recommended that anyone proposing to release transgenic organisms should provide a worst case scenario so that possible impacts could be fully assessed in advance (see Lee 1992a, p. xxvi). Clearly, this was not done in the case of the gene-bean.

As it stands, ANZFA's proposed standard will not adequately protect public health, safety, or the consumer's right to know. Though foods produced using gene technology have no history of safe use, there is no proposal to undertake pre-market human testing or post-market monitoring. Without such mechanisms, general release of novel products into the food supply would further make all Australians and New Zealanders part of the giant uncontrolled r-DNA experiment. The potential for new genes and proteins to create allergic reactions or other illnesses is poorly understood. Yet, after short trial periods, whole categories of food could be registered for general commercial release.

Foods would be assessed and approved case-by-case but assessments would be based on the concept that most genetically-manipulated foods are 'substantially equivalent' to their conventional counterparts. This approach focuses on the superficial similarities of end products rather than the basic, potentially dangerous, differences between conventional and r-DNA production processes and their effects. Substantial equivalence is a concept dreamed up by industry to support the fantasy that genetic engineering is not radically different from traditional breeding. Yet, in traditional breeding the transfer of genetic material between unrelated or distantly-related species is nigh impossible. As a rational and sensible basis for labelling policy, all transgenic food and foods containing transgenic material should be labelled so that shoppers can make informed choices at the supermarket and drugstore.

At a Food Summit convened by the Federal government in 1996, the broad spectrum of community groups represented all supported the development of an adverse reactions register, to record any illness or allergy caused by novel foods. This would alert health authorities to any pattern of harm that may arise. It is, however, not in the novel food standard proposed by ANZFA, which argues that ANZFA's assessment strategy ensures prevention, making a register unnecessary. But things can go wrong, especially as foods do not undergo the kinds of rigorous testing required of drugs. This was potently and tragically signalled in 1989 when a batch of the food supplement L-tryptophan manufactured using genetically-engineered microbes (Strain V) entered US and European markets from Japan. It killed at least 37 people and permanently crippled thousands more with a new disease, the central nervous system disorder—eosinophilia myalgia syndrome (EMS) (Mayeno and Gleich 1994). Months went by before the culprit was detected because, among other things, the product bore no label signalling its radical process of manufacture.

Manufacturer Showa Denko has since settled million-dollar damages claims out of court, but without being required legally to produce for analysis the organism that caused the havoc. In extreme damage-control mode, the gene technology industry still argues that changes in the purification process during manufacture of the potent L-tryptophan batch may have been responsible, playing down the eventuality that the toxin was produced by Strain V (Roberts 1990; Troutwine 1991; Swinbanks and Anderson 1992). Unless its assessment processes are greatly strengthened and democratised, ANZFA's standard might allow a similar opportunity for potentially toxic r-DNA foods to slip through, be registered, and sold without a label.

Under ANZFA's proposal no general labelling of r-DNA foods would be required. This is the more disconcerting because public opinion surveys indicate that the majority of Australians want the process of production on labels (Kelley 1995). The support for caution and labelling is also reflected in thousands of submissions on Proposal P97 to ANZFA calling for labels, and thousands of petition signatures to the Senate asking for labelling and per-market human testing. Even if the ingredient list includes the term 'genetically altered', ANZFA proposes that the label could also claim 'genetically enhanced, improved or augmented', despite the reduced quality and safety posed by foods like the gene-beans. In an example of the proposal in action, ANZFA says a sweeter tomato will be labelled because the end product is different. But without a sound rationale, foods from plants engineered with such things as Bt toxins (insect killer), viral particles (virus resistant), antibiotic resistance marker genes (in most engineered plants as part of the production process), and herbicide tolerance (leaving up to 200 times the present level of chemical residues), will be regarded as 'substantially equivalent' to conventional plants and will not be labelled. This amounts to an institutional bias in favour of r-DNA food manufacturers, a bias which would obviously act to negate, misrepresent, or convert established public opinion. (As Hindmarsh argues in chapter 3, such strategies are designed to *manufacture consent*.)

Enzymes, additives, hormones and processing aids produced by genetically-engineered microorganisms (as with L-tryptophan) would not be labelled. The cheese rennet, baking and brewing enzymes, and synthetic pig growth hormone, which are already in processed foods, would remain unlabelled. To achieve the best outcomes for the Australian public, ANZFA's rules must become more responsive to the many legitimate concerns of citizens. Strong regulation, assessment, monitoring, real public participation, and universal labelling would go some way to rebuilding the confidence and trust that our food industry requires if it is to restore its increasingly tarnished credibility.

SUSTAINABLE AGRICULTURAL SYSTEMS

To achieve the essential goals of diversity and sustainability in agriculture, the GeneEthics Network argues that 'adequate government funding should be provided to encourage research into alternatives to genetic engineering technologies' (ACF 1989). Sustainable systems such as organic farming and agroecology that work harmoniously with ecological processes are hampered by minimal research funding, few tax breaks, and lack of official support. Research into sustainable agriculture, both in Australia and overseas, receives less than one per cent of the government funding allocated to agricultural research and development.

Though the CSIRO Division of Land and Water recognises the environmental problems created by European industrial agriculture in Australia, its current $60 million, five-year research project to overhaul agriculture will chiefly explore high technology futures, at the expense of more diversified options (Collis 1997). The high-tech quick-fix approach primarily addresses the symptoms of environmental degradation and ignores the causes of the problems, which are addressed by the whole systems approach of organic agriculture. Quick-fixes will not save the nation's ecosystems now being depleted at a rate far beyond replacement (Phelps 1997). The desertification, salinity, soil loss and water pollution created by present rural management practices and technologies will most likely be extended, and entrenched, through the use of ever more marginal and fragile ecosystems for intensive production using gene technology products. This would be the likely case, for example, were plants designed to grow in saline and arid zones—as is occurring with current genetics experimentation (see chapter 11).

Genetic engineering is by no means the solution to the many social and environmental ills that our planet now suffers, as the genetic engineers and their industrial patrons would have us believe. They sell the technology to policy makers as a future money-spinner and 'cure-all' to receive increasing levels of agricultural research resources. Meanwhile, alternative, sustainable agricultural systems remain underfunded or have their meagre funding cut. As most public sector science is now required to have business support, funds go to technologies that are tied to patentable and profitable outcomes for capital accumulation, not to those that will produce the best integrated environmental, social and economic outcomes. Benign agricultural management systems such as organics, herbal medicines, and open-pollinated crop varieties are marginalised and unfunded because they cannot be easily monopolised by big business interests.

In a further strategy to expand biotechnology markets and undermine the competition from increasingly popular organic foods, in North America, Europe, and in the international *Codex Alimentarius* Committee meetings, the food and fibre industries are lobbying hard to have genetically-engineered organisms included in the organic standard (Burnstein 1997; Pure Food Campaign 1997b). If this were eventually successful, pressure for Australia to follow suit would be based on staying in line with overseas standards and being competitive in global markets.

In his report *Priority Matters*, Australia's Chief Scientist John Stocker proposes that national goals rather than the lobbying ability of scientists and special interests should set this nations R and D priorities (see Stocker 1997). Comprehensive, informed public debate is thus needed to challenge the dominance of technical experts, and to facilitate community-wide reappraisal and implementation of our priorities. Democratic, community processes of decision making are the only way to arrive at truly sustainable systems for the twenty-first century.

Contemporary Australian society urgently needs to repair the disasters of industrial agriculture. Yet little notice is being taken of the potential of genetic engineering to worsen such disasters. This is because the adoption of genetic engineering lies mainly in the interests of corporate capital. Gene technology is being adopted by technocratically-dominated and myopic governments as the 'bio-future' for agricultural production, without public participation, consultation or acceptance.

CITIZENS' CAMPAIGNS AGAINST GENETIC ENGINEERING

The Australian Conservation Foundation (ACF) began its genetic-engineering campaign in 1988, with royalties from the Midnight Oil band's album *Species Deceases*. In 1991, the GeneEthics Network was formed, with a federal government grant to support and continue the campaign. Genuine public control over all aspects of genetic engineering and its applications—to replace the token consultation and marginal representation by critics and the public—is the Network's goal. With the Liberal–National Party Coalition taking office at the 1996 federal election, the Network's official support ended abruptly and it now relies solely on the generosity of its growing contingent of supporters.

The Network is engaged in a range of lobbying, activist and networking activities. These include policy analysis and development, lobbying policy makers and government departments, conducting public awareness and education campaigns, liaising with the news media, developing education and campaign materials, organising and participating in conferences and public debates, linking activists around campaigns of common interest, and staging protest initiatives.

One of the Network's central goals is independent assessment and regulation of genetically-engineered organisms intended for release into the environment. Network members always argue that the body responsible for these functions should be separate from those organisations promoting and funding the research and development of biotechnology. It is a scandal that GMAC is located within the Gene Technology Section of the Department of Industry and Science, and is dominated by bioscientists. Social, ethical and economic costs and benefits are ignored.

The Network has played a leading role in disseminating information, to enable citizens to engage in informed debate over whether they want a bio-future which includes gene technology and its products. When transgenic pigs were released into Adelaide's food supply without authorisation or public knowledge in 1988, the Network was eventually able—with inside information—to break the attempted 'cover-up' in the national news media (ACF 1990). In 1992, public awareness and discussion forums were held in capital cities. The *Gene File* newsletter and the Network's HomePage also offer a range of information and educational materials. The Network has published *The Troubled Helix*—a widely used set of resources and case studies for high schools and universities, on the environmental, social, and ethical issues of gene technology.

Networking between groups, to help develop common campaigns on issues of mutual concern has also been a key feature of the group's work. Other organisations working with the Network include: Fitzroy's Friends of the Earth Anti-Genetic Engineering Collective which focuses on environment and social justice questions; the Consumers Federation of Australia Food Group which develops policy on regulation of the food supply; EcoConsumer which researches and publicises industry programmes and policies, and the Australian Consumer's Association which also lobbies government and attempts to raise public awareness.

The Network also liaises with groups actively engaged in sustainable agriculture or which support the replacement of industrial agriculture with sustainable systems—NASAA, ORGAA, the Biodynamic Research Institute, Permaculture, Biological Farmers of Australia, the Diggers' Club, the Heritage Seed Curator's Association, the Seed Savers Network, as well as with allergy and sensitivity groups, feminist groups, doctors, natural therapists, animal rights groups, concerned scientists, and many others. The Network has built extensive links with various overseas organisations that campaign strongly against genetic engineering, including: Greenpeace International (Holland), No Patents on Life (Germany), the Pure Food Campaign (USA), the Third World Network (Malaysia), the Research Foundation for Science, Technology and Ecology (India), the Biotechnology Working Group (USA), the Genetics Forum (England), FINRRAGE (Germany), and the Rural Advancement Foundation International (Canada). These and many other groups organised the Global Days of Action in 30 countries during April and October 1997, where the focus of activity was opposition to genetically engineered foods, factory farming, and life patents.

Also at the international level, the GeneEthics Network's activities have included lobbying Australian diplomats during the development of a Biosafety Protocol to regulate transnational movements of genetically-modified organisms, and international Codex standards on the labelling of genetically-engineered foods. In both of these forums, the Australian government is unfortunately siding with the big biotechnology players of the OECD to ensure that regulations and standards are as weak as possible (see Australian GeneEthics Network 1997, pp. 3, 5).

CONCLUSION

The preferred future of the Australian GeneEthics Network is a stable, just, and equitable society based on local—nature-friendly—sustainable systems of production and consumption. Citizen support worldwide is growing fast for participatory technology assessment and government intervention to make democratic processes really work to control the unbridled power of corporations. There are still opportunities to leave a good legacy for future generations—a world free of genetic pollution, the private ownership of life, arrogant notions of reconstructing Nature, social control and eugenic abuses based on simplistic theories of genetic-determinism, or of biotechnical elites who continue to make important decisions without democratic participation or citizen choice. We reject the proponents' plan for what they project as a genetically-engineered Utopia.

Glossary

Entries have been drawn directly from five main sources. We would like to thank the publishers of those sources for giving us permission to use the entries cited, including: Oxford University Press (*A Concise Dictionary of Biology*, 1990, 2nd edn; Stoddart Publishing Co. Ltd, Canada (Suzuki, D. and Knudston, P., 1988, *Genethics*); Zed Books (Shiva, V. and Moser, I., 1995, *Biopolitics*); Pan Macmillan Australia (Haynes, R., 1991, *High Tech: High Co$t*); Pluto Press (Wheale, P. and McNally, R., 1990, *The Bio-Revolution: Cornucopia or Pandora's Box*). Where applicable, the key for entries is listed at the end of the glossary.

agrochemical Synthetic chemical used in industrialised agriculture for plant production (fertiliser) or protection (pesticide) (1).

allergy A condition in which the body produces an abnormal immune response to certain antigens (called *allergens*), which include dust, pollen, certain foods and drugs, or fur. In allergic individuals these substances, which in a normal person would be destroyed by antibodies, stimulate the release of histamine, leading to inflammation and other characteristic symptoms of the allergy (for example, asthma or hay fever) (+5).

amino acids The base chemical subunit or building blocks of proteins; there are twenty common amino acids (+3).

bacillus A rod-shaped bacterium.

Bhopal The world's worst chemical accident. On 2 December 1984 noxious fumes containing methyl-isocyanate, hydrogen cyanide, and cyanide leaked out from the Union Carbide plant in Bhopal, India. Up to 16 000 people are thought to have been killed as a result of breathing the toxic fumes, another 125 000 suffered some health damage—at least 30 000 of whom are described as incurably ill. Union Carbide settled for US$470 million in 1991 and quietly left the country (see Rowell 1996, pp. 96–7).

biochemistry The study of the chemistry of living organisms, particularly the structure and function of their chemical components (principally proteins, carbohydrates, lipids and nucleic acids) (+5).

biodevelopment Bio-industrial development.

biodiversity The millions of life forms found on earth, their genetic variety and variability, the ecological roles they perform and the interrelated and interdependent ecological communities or ecosystems they form, or are found in (+1).

bioindustrial Industries engaged in the life sciences.

biology The study of living organisms, which includes their structure (gross and microscopial), functioning, origin and evolution, classification, interrelationships and distribution (5).

biosciences The biological or life sciences include biochemistry, biomedicine, cellular biology, chemical engineering, microbiology, molecular biology.

bioscientists Scientists engaged in research and development in the life sciences.

biotechnology Development of products by exploiting biological processes or substances. Production may be carried out by using intact original or modified organisms, such as yeasts and bacteria, or by using active cell components, such as enzymes from organisms; the attempt to engineer and control biological processes for the purposes of industrial production (1).

bio-Utopia Is short for biological Utopia; the Enlightenment scientist and philosopher Francis Bacon (1561–1626) is credited with advancing the 'scientific' notion of a biological utopia in his book *New Atlantis* (see, for example, Weinberger 1989, pp. 73–5; Krimsky 1991, pp. 84–5). His writings advocate the creation of new species (or new 'strains' of species) as liberating an 'oppressed' humanity from hunger, disease etc.—just as genetic engineers now, some 300 years later, advocate the purpose of genetic engineering. The morality of creating new species is primarily anthropocentric, or human-centred, that is, Nature exists to 'serve' humans. In the extreme form of this morality, the environmental ethics of gene-altering are ignored, denied, or pushed aside.

blood plasma The liquid part of the blood excluding blood cells. It comprises water containing a large number of dissolved substances, including proteins, salts, food materials (glucose, amino acids, fats), hormones, vitamins, and excretory materials (+5).

Down's syndrome A congenital form of mental retardation due to a chromosome defect (+5).

bovine somatotropin (bST) A commercially manufactured form—using genetically engineered microbes—of bovine growth hormone produced in the pituitary glands of cows essential to milk production.

Regular injections of biosynthetic BGH—known as bST (or r-bST)—can raise the productivity of dairy herds. Use of r-bST is presently prohibited in Australia and the European Community, but is common in the USA. Anti-bST campaigners claim that its use is cruel because of the frequency of injections and increased risks of mastitis in cows, that it poses a risk to human health both as a possible carcinogen and through over-use of antibiotics (to treat mastitis), and that it represents an economic threat to small-scale dairy farms.

BRCA1 Breast cancer gene 1. Located on chromosome 17. The link between early-onset breast cancer to BRCA1 gene (for breast cancer gene 1) was first discovered by Hall *et al.* (1990). Soon the discovery of two other mutations, one called p53 (Malkin *et al.* 1990), the other BRCA2 (Wooster *et al.* 1994), on chromosome 13, followed. Women who inherit the mutated form of the gene have a substantially increased lifetime risk of developing breast or ovarian cancer.

carrier In some genetic mutations, such as cystic fibrosis, which affects about one in 2500 Australians, the genetic defect is unnoticeable in a carrier. In carriers only one of the two inherited copies of the gene is defective. The defect is masked by the second healthy copy, and such individuals do not suffer from the condition themselves. However, they can pass the defective copy on to their offspring. Thus, if two carriers produce children, each child has a one in four chance of being afflicted with the disease.

cell The structural and functional unit of all living organisms (+5).

cell culture Growth of cells under laboratory conditions.

central dogma The belief of molecular biology that hereditary information generally flows unidirectionally from DNA molecules to RNA molecules to proteins (see also genetic determinism).

CFTR Cystic fibrosis transmembrane conductance regulator gene; the genetic defect in cystic fibrosis lies in this gene. It has a role in the transport of ions; more than 400 mutations in the gene have been reported to be responsible for the dysfunctions, thus accounting for the striking variation in lung function between afflicted individuals, although other genetic and non-genetic factors, as yet not identified, may also influence the manifestation and prognosis of the disease.

chromosome The condensed rod made up of a linear thread of DNA interwoven with protein that is the gene-bearing structure of eukaryotic cells (3).

clone A group of genetically identical cells or organisms produced asexually from a common ancestor; all cells in the clone have the same genetic material and are exact copies of the original (1).

cloning The process of asexually producing a group of cells (clones), all genetically identical, from a single ancestor; in recombinant-DNA technology, the use of various procedures to produce multiple copies of a single gene or segment of DNA is referred to as 'cloning DNA' (4).

containment (physical) Measures designed to prevent or minimise the escape of recombinant organisms (+2).

cytogenetics The study of inheritance in relation to the structure and function of cells (+5).

cytoplasm The part of the cell that lies between the membrane and the nucleus. It contains a variety of subcellular structures that participate in the cell's functions (4).

DNA (deoxyribonucleic acid) The self-replicating molecule that is the carrier of genetic information in nearly all living organisms. Every inherited characteristic has its origin somewhere in the code of each individual's DNA (1).

double helix The name given to the structure of the DNA molecule, which is composed of two complementary strands which lie alongside and twine around each other, joined by cross-linkages between base pairs—two nucleotide bases on different strands of the nucleic acid molecule that bond together (+2).

ecologically sustainable development (ESD) Types of social, economic and technological development which sustain the natural environment and promote social equity.

ecosystem A community of life forms, interacting with each other, together with the environment in which they live and with which they also interact (1).

enzyme Any one of a large class of proteins occurring in organisms which act as catalysts, that is, make it possible for chemical reactions to occur and to occur sufficiently rapidly to meet the organism's needs (4).

***Escherichia coli* (*E.coli*)** A bacterial species that inhabits the intestinal tract of most vertebrates and on which much genetic work has been done; some strains are pathogenic to humans and other animals; many non-pathogenic strains are used experimentally as hosts for recombinant-DNA (2).

eugenics A strategy of trying to orchestrate human evolution through programmes aimed at encouraging the transmission of 'desirable' traits and discouraging the transmission of 'undesirable' ones (3).

gene The basic physical and functional unit of heredity that is transmitted from one generation to the next and can be transcribed into a polypeptide or protein (3).

gene mapping Determination of the relative positions of genes on a chromosome.

gene therapy The application of genetic engineering techniques to alter or replace 'defective' genes. The techniques are still at the experimental stage (+5).

genetic determinism The theory that the phenotype (the detectable characteristics associated with a particular genotype) is an innate and essentially unchangeable expression of the genotype (+2). Genetic determinism (or essentialism—see chapter 9) is now also being referred to as *geneticisation*—'an ongoing process by which differences between individuals are reduced to their DNA codes, with most disorders, behaviors and physiological variations defined, at least in part, as genetic in origin' (Lippman 1991 cited in Hubbard and Wald 1997, p. 187). A more general statement, based upon a critical perspective, is this: 'The new biotechnology based upon genetic engineering makes the assumption that each specific feature of an organism is encoded in one or a few specific, stable genes, so that the transfer of these genes results in the transfer of a discrete feature. This extreme form of genetic reductionism has already been rejected by the majority of biologists and many other members of the intellectual community because it fails to take into account the complex interactions between genes and their cellular, extra-cellular and external environments that are involved in the development of all traits.' (Third World Network 1995, p. 10)

genetic engineering The technique of altering the genetic make-up of cells or individual organisms by deliberately inserting, removing or altering individual genes (+3).

genetic monitoring Tests that examine hereditary molecules for early indications of genetic damage or disease (3).

genetics The branch of biology concerned with the study of heredity and variation (+5).

genetic screening The process by which individuals are asked to undergo genetic testing; the process of systematically scanning individual genotypes for possible hereditary defects or abnormalities (3).

genetic testing Analysis of human DNA, RNA, chromosomes, or gene products for early indications of genetic damage or disease.

genome The entire genetic endowment of an organism or individual (3).

genotype An individual's genetic make-up underlying a specific trait or constellation of traits (3).

GEO Genetically engineered organism.

germ cell Reproductive cell.

Guthrie test A biochemical test, not one based on molecular (genetic) technology.

hybrid A molecule, cell or organism produced by combining the genetic material of genetically dissimilar organisms; traditionally, hybrids were produced by interbreeding whole animals or plants; cell fusion technology and transgenic engineering are innovations in hybridisation (2).

hybridoma A 'hybrid myeloma'; a cell produced by the fusion of an antibody-producing cell (lymphocyte) with a cancer cell (myeloma) (2).

hybridoma technology The technology of fusing antibody-producing cells with tumour cells to produce in hybridomas which proliferate continuously and produce monoclonal antibodies (2).

immunology The study of phenomena related to the body's response to antigenic challenge (that is, immunity, sensitivity, and allergy).

intellectual property rights Laws that grant monopoly rights to those who create ideas or knowledge. They are intended to protect inventors against losing control of their ideas or the creations of their knowledge. There are five major forms of intellectual property: patents, plant breeder's rights, copyright, trademarks, and trade secrets. All operate by exclusion, and grant temporary monopolies which prevent others from making or using the creation. Intellectual property legislation is national, although most countries adhere to international conventions governing intellectual property, and all members of the World Trade Organisation must have intellectual property rights covering living organisms. Patents and Plant Breeder's Rights are the forms most relevant to living organisms.

in vitro: Biological reactions occurring outside of the living organism, in test tubes or any laboratory containers—all of which are artificial systems. Latin term meaning 'in glass'.

***in vitro* fertilisation (IVF)** The fertilisation of an egg cell by sperm on a glass dish (2); fertilisation carried out in the laboratory, outside a woman's body (1).

Janus-faced two-sided; Janus was an ancient Roman god—the guardian of doors and gates, and who was represented with faces on the front and back of the head.

messenger RNA An RNA molecule that has been transcribed from a gene-bearing DNA molecule and will later be translated into a protein (3).

microbiology The study of living organisms that can only be seen under a microscope.

microorganism (microbe) An organism that can only be seen with the assistance of a microscope.

molecular biology The study of the structure and function of large molecules associated with living organisms, in particular proteins and the nucleic acids DNA and RNA (5).

molecular genetics The study of the molecular basis of gene structure and function (3).

monoclonal antibody One of a clone of antibodies produced by a hybridoma (2).

monoculture The agricultural practice of cultivating crops consisting of genetically similar organisms (3).

mutation A heritable change in a DNA molecule (3).

neo-luddite: Person actively opposed to the introduction of (a) new technology; after the Luddites, an organised band of mechanics which went about destroying machinery in the midlands and north of England in the early nineteenth century; derived from Ned Lud, who, in the latter half of the eighteenth century, smashed up machinery belonging to a Leicestershire manufacturer as a protest against mechanisation (2).

nucleic acid One of a number of acidic, linear, information-bearing, biological molecules belonging to one of two chemical families—DNA or RNA (3).

nucleotide A chemical subunit composed of a sugar, phosphate and base, which makes up the nucleic acids DNA and RNA (3).

organism Any form of living entity, whether plant, animal, bacterium, protist or fungus.

patents Legal monopolies that cover a wide range of products and processes, including life forms. To be patentable, inventions must theoretically meet three basic criteria. They must be: (a) novel—they must not have been known previously to the public; (b) useful—they must do what they claim, though they need not necessarily be practical; and (c) non-obvious—they must have an 'inventive step' and constitute some notable extension of what was previously known. Patents provide exclusive legal protection to patent holders, usually for seventeen to twenty-five years. Anyone wishing to use a patented invention must receive permission from the patent holder and often must pay a royalty. In exchange for this monopoly, the patent holder must disclose or describe the invention.

pathogen A microorganism, such as a bacterium or virus, that infects another organism to produce disease symptoms or a toxic response (1).

plant breeder's rights (plant variety rights) A form of intellectual property law that grants a plant breeder's certificate to those who breed new plant varieties. Plant breeder's rights generally contain a breeder's and research exemption that allow non-commercial use of protected varieties. In the US, recent court decisions have threatened these exemptions. There are currently two international agreements governing Plant Breeder's Rights, both of them under UPOV—the International Convention for the Protection of New Plant Varieties.

plasmid A circular, self-replicating form of DNA found in many species of bacteria that can sometimes be used as a vector to ferry recombinant genes into another species (3).

polygenic Controlled by or associated with more than one gene (3).

polypeptide A protein or portion of a protein consisting of two or more amino acids (3).

population A local group of organisms belonging to the same species and capable of interbreeding (3).

probe (molecular) A DNA or RNA molecule, labelled with a radioactive isotope, dye, or enzyme, used to locate a specific nucleotide sequence or gene on a DNA molecule.

protein A large molecule composed of one or more chains of amino acids in a specific sequence; the sequence is determined by the sequence of the nucleotides in the gene coding for the protein; proteins are the doing molecules of the cells; they are required for the structure, function, and regulation of the body's cells, tissues, and organs, and each protein has unique functions (4).

recombinant-DNA (r-DNA) A novel DNA sequence produced by artificially joining pieces of DNA from different organisms together in the laboratory (3).

reductionism The philosophical belief that phenomena or organisms are best understood by breaking them up into smaller parts (4); a belief that all phenomena find their ultimate explanation in terms of elementary physico-chemical processes and events occurring at the level of atoms and molecules (see also genetic determinism above).

reproductive technology Largely experimental procedures which involve medical intervention in the process of conception (1).

ribosomes The particles in the cell cytoplasm on which the base sequences brought there by messenger-RNA get translated into the amino acid sequence of proteins (4).

RNA (ribonucleic acid) A class of nucleic acids that closely resembles DNA but has ribose sugars rather than deoxyribose sugar and uracil rather than thymine bases (3).

sequence The order of nucleotides in a nucleic acid or the order of amino acids in a protein (4).

sociology The study of society. The two principle objects of study are social facts and social processes.

somatic cell A body cell that is not destined to become an egg or a sperm (3).

species A unit of biological classification; sexually reproducing organisms are classified as belonging to the same species if they can interbreed and produce fertile offspring; the interpretation of what constitutes a species is controversial; it is especially difficult to apply the concept of species to bacteria which are not subject to reproductive isolation (2). In microbiology, variants within species are called strains.

systems A system is a set of variables so interconnected that a change in the value of one of the variables has a determinate effect on all other variables (Mann 1983, p. 388).

Three Mile Island A nuclear reactor accident near Harrisburg, Pennsylvania in which water flow in the reactor primary circuit was interrupted, leading to increased pressure, chemical reaction and partial melting of the reactor core. Radioactivity release was minimal but the cost of decommissioning the reactor exceeded $4 billion (+1).

tissue culture *In vitro* growth in nutrient medium of cells isolated from tissue.

toxin A substance, in some cases produced by disease-causing microorganisms, that is poisonous to living organisms (2).

transgenic An organism produced by the transfer and expression of genetic material (DNA) derived from a different species.

vaccine A preparation of killed or debilitated microorganisms—or their components or products—used to induce immunity against a disease.

Key for entries

(1) *High Tech: High Co$t*
(2) *The Bio-Revolution*
(3) *Genethics*
(4) *Biopolitics*
(5) *A Concise Dictionary of Biology*

+ symbolises a shortened or sightly amended version of the cited entry.

Bibliography

Abbott, A. (1996) 'Transgenic Trials under Pressure in Germany', *Nature*, 14 March, p. 94

Agbiotech Bulletin (1997) 'Competitiveness of Biotechnology in Europe', 5, 10, p. 8

Alexander, N. (1992a) 'Will Pigs Fly? Taking Genetic Engineering to the People', *Search*, 23, pp. 210–11

——(1992b) (Manager, CSIRO Film and Video Centre), *pers comm.*, 22 May

AMRAD (1995) AMRAD and Northern Land Council Sign Agreement for Pharmaceutical Screening on Arnhem Land Plants, AMRAD News Release, 19 July

——(1996) *Annual Report*, Melbourne

——(1997a) AMRAD Announces Natural Products Screening Collaboration, AMRAD News Release, Melbourne, 9 January

——(1997b) AMRAD Enters Multi-million Dollar Natural Products Screening Collaboration with Rhone-Poulenc Rorer, AMRAD News Release, Melbourne, 12 May

Anderson, I. (1997) 'Viral Renegades', *New Scientist*, 6 September, p. 10

Anderson, J., Moore, J. and Hill, R. (1997) New Regulations for Gene Technology, Joint Press Release, DIST, Canberra

Armstrong, J. and Hooper, K. (1994) 'Nature's Medicine', *Landscape* 9, 4, pp. 12–15

Australian 10 August 1976; 19 March 1994; 8 August 1994; 23 May 1995; 19–20 October 1996; 15 October 1997

Australian Biotechnology Association (ABA) (1993) *Australian and New Zealand Biotechnology Directory 1993*, 3rd edn, Australian Biotechnology Association Ltd, Melbourne

——(1996a) Regulation of Genetic Engineering, http://www.aba.asn.au/leaf7.html

——(1996b) Biotechnology in animal agriculture, http://www.aba.asn.au/leaf4.html

Australasian Biotechnology (1991) 1, 1

——(1996) 'Company News', 6, 6, p. 335

Australian Academy of Science, (1980) *Recombinant DNA: An Australian Perspective*, Australian Academy of Science, Canberra

Australian Broadcasting Commission (ABC), *Four Corners*, 18 March 1977, 11 September 1989
Australian Conservation Foundation (ACF) (1989) Policy Statement Number 45, Genetic Engineering, 9 October, ACF, Melbourne
——(1989) Genetic Engineering Campaign Report to ACF Council, Melbourne, December
——(1990) Genetically Engineered Pigs Scandal Covered Up, News Media Release, 26 April
——(1992) *Genetic Engineering: Science Before It's Time*, ACF, Melbourne
Australian Cotton Outlook (1997) January
Australian Food Council (1996) Food Biotechnology in Australia: Issues Relating to the Application of Gene Technology, Discussion Paper, Australian Food Council, Canberra
Australian GeneEthics Network (1997) *The Gene File*, 1, Australian GeneEthics Network, Melbourne
Australian Industrial Property Organisation (1997) Australian Patents for: Micro-organisms, Cell Lines, Hybridomas, Related Biological Materials and Their Use, Genetically Manipulated Organisms, AIPO, Canberra
Australian Medical Association (AMA) (1995) *Revised Code of Ethics*, Australian Medical Association, Barton, ACT
Australian Research Council (ARC) (1994) Access to Australia's Genetic Resources, Draft Report of the Australian Research Council Workshop
Australian Science and Technology Council (ASTEC) (1982) *Biotechnology in Australia*, Australian Government Publishing Service, Canberra
Avcare (1995) *Facts and Figures*, 3rd edn, Avcare (National Association for Crop Protection and Animal Health), Sydney
Bak, P. (1996) *How Nature Works*, Springer Verlag
Barbano, D. (1997) Bovine Growth Hormone: A Cornell University bST Fact Sheet, http://www.babybag.com/articles/bovine.htm
Bartels, D. (1984) 'Secrecy in Biotechnology is Shortsighted', *Search*, 15, pp. 7–8
Beck, U. (1992) *Risk Society: Towards a New Modernity*, Sage, London
——(1995) *Ecological Enlightenment: Essays on the Politics of the Risk Society*, Humanities Press, New Jersey
——(1996) 'Risk Society and the Provident State', in Beck, U., Giddens, A. and Lash, S. (eds) *Reflexive Modernization*, Polity Press, Cambridge
Beck-Gernsheim, E. (1996) 'Life as a Planning Project', in Lash, S., Szerszynsky, B. and Wynne, B. (eds) *Risk, Environment and Modernity*, Sage, London
Bernstein, P. (1996) *Against the Gods: The Remarkable Story of Risk*, Wiley and Sons, New York
Bishop, R. (1974) 'Anxiety and Readership of Health Information', *Journalism Quarterly*, 51, 1, pp. 40–6
Bolt, C. (1997) 'Japan "Nervous" of Transgenic Crops', *The Financial Review*, 16 June, p. 7
Britz, M. (1991) 'Food Biotechnology: the Industry and the Issues', *Australasian Biotechnology*, 1, 1, pp. 30–2
Brough, E., Foster, J., Norton., G. and Holdom, D. (1995) Integrating New Biotechnology in Crop Protection, Paper Presented at the *4th Pacific Rim Biotechnology Conference*, Melbourne, February

Brown, D. (1997) Farmers to Fight 'Scare Stories' on Super Crops, *Daily Telegraph*, 13 January

Brown, M. (1995) Lazarus Bacteria Shows Signs of Life after 30 million Years, *Sydney Morning Herald*, 20 May, p. 1

Brumfield, J. (1980) Science Rejects DNA Fears, *The Australian*, 30 July, p. 9

Brungs, R. (1983) 'The Human Body: Artefact or Icon?', in Santamaria, J. and Tonti-Filippini, N. (eds) *Proceedings of the 1983 Annual Conference of St Vincent's Bioethics Centre*, St Vincent's Bioethics Centre, Melbourne

Bruno, K. (1997) 'Say it Ain't Soy, Monsanto', *Multinational Monitor*, Jan–Feb, pp. 27–30

Bunyard, P. (1996) 'Industrial Agriculture-Driving Climate Change?', *The Ecologist*, 26, 6, pp. 290–98

Burch, D., Hulsman, K., Hindmarsh, R. and Brownlea, A. (1990) Biotechnology Policy and Industry Regulation: Some Ecological, Social and Legal Considerations, Submission to the House of Representatives Standing Committee of Industry, Science and Technology Inquiry into Genetically Modified Organisms, September

Burch, D., Rickson, R. and Annels, R. (1992) 'The Growth of Agribusiness: Environmental and Social Implications of Contract Farming', in Lawrence, G., Vanclay, F. and Furze, B. (eds) *Agriculture, Environment and Society: Contemporary Issues for Australia*, Macmillan, Melbourne

Burch, D., Rickson, R. and Lawrence, G. (eds) (1996) *Globalization and Agrifood Restructuring: Perspectives from the Australasia Region*, Avebury, London

Bureau of Rural Resources (1991) Biotechnology in Australia: Perspectives and Issues for Animal Production, Working Paper WP/16/91, Bureau of Rural Resources, Canberra

Burnstein, S. (1997) 'Organics Under Siege', *Food and Water Journal*, Winter, pp. 20–1

Busch, L. (1991) 'Biotechnology: Consumer Concerns About Risks and Values', *Food Technology*, April, pp. 96–101

——, Lacy, W., Burkhardt, J. and Lacy, L. (1991) *Plants, Power and Profit: Social, Economic and Ethical Consequences of the New Biotechnologies*, Basil Blackwell, Massachusetts

Butler, D. (1997) 'Bioethics Needs Better Input From Public', *Nature* 389, 6653, p. 775

Campus Review (1997) 23–29 July

Carey, A. (1995) *Taking the Risk Out of Democracy*, University of New South Wales Press, Sydney

Carey, J. (1989) *Communication as Culture. Essays on Media and Society,* Routledge, New York

CC (Cotton Conference) (1996) *Cotton On To The Future: Proceedings of The Eighth Australian Cotton Conference*, Conrad Jupiters, Gold Coast, 14–16 August

Charlesworth, M., Farrall, L., Stokes, T. and Turnbull, D. (1989) *Life Among the Scientists*, Oxford University Press, Melbourne

CSIRO (1981) Biotechnology For Australia: Report to the Executive of CSIRO, June

Clark, N. (1997) 'Panic Ecology: Nature in the Age of Superconductivity', *Theory, Culture and Society*, 14, 1, pp. 77–96

Collis, B. (1997) Reinventing Aussie Farming, *Canberra Times*, 16 September, p. 14

Connor, S. (1997) Scientists Find Gene For Intelligence, *London Sunday Times*, 19 October

CornerHouse (1997) *No Patents on Life!*, The CornerHouse, Dorset, UK, September

Cotton Research and Development Corporation (CRDC) (1997) The Performance of INGARD Cotton in Australia 1996/97 Season, Occasional Paper, CRDC, Narrabri

Couchman, P. and Fink-Jensen, K. (1990) Public Attitudes to Genetic Engineering in New Zealand, DSIR, Christchurch

Cribb, J. (1994) The Next Generation, *Australian*, 10 August, p. 9

Crick, F (1958) 'On Protein Synthesis', *Symp. Soc. Exp. Biol.*, 12, pp. 138–63

——(1970) 'Central Dogma of Molecular Biology', *Nature*, 227, pp. 561–63

Crook, S. (1997) Ordering Risks, Paper Presented to Risk and Socio-cultural Theory Conference, Centre for Cultural Risk Analysis, Charles Sturt University, Sydney, April

Daniels, K. and Murnane, M. (1980) *Uphill All The Way: A Documentary History of Women in Australia,* University of Queensland Press, Brisbane

Danks, D. (1993) 'Whither Genetic Services?', *The Medical Journal of Australia*, 159, pp. 221–22

Davis, A. (1984) Ethical Issues in Prenatal Diagnosis, *British Medical Journal*, 288

Dawkins, K. (1997) Institute for Agriculture and Trade Policy, correspondence to US Secretary of State Madeleine Albright, 30 June

Dawkins, R. (1986) *The Blind Watchmaker*, Longman Scientific and Technical, Harlow

Decima Research (1993) Final Report to the Canadian Institute of Biotechnology on Public Attitudes towards Biotechnology, Canadian Institution of Biotechnology, Ottawa

Dessaix, R. (1994) Another Country, *24 Hours*, May 1994, p. 148

Douglas, M. (1966) *Purity and Danger*, Routledge and Kegan Paul, London

——(1983) 'Environments at Risk', in Barnes, B. and Edge, D. (eds) *Science in Context: Readings in the Sociology of Science*, Open University Press, Milton Keynes

——and Wildavsky, A. (1982) *Risk and Culture*, University of California Press, Berkeley

Dow, S. (1997) Gene Patent Decision Under Fire, *The Age*, 10 June

Draper, E. (1991) *Risky Business: Genetic Testing and Exclusionary Practices in the Hazardous Workplace*, Cambridge University Press, New York

Dredge, R. (1990) Case of the Super Pig that Went to Market, *The Age*, 2 May

Dunwoody, S. (1993) *Reconstructing Science for Public Consumption: Journalism as Science Education*, Deakin University, Victoria

East Asia Analytical Unit (1994) Subsistence to Supermarket: Food and Agricultural Transformation in South-East Asia, Department of Foreign Affairs and Trade, Australian Government Publishing Service, Canberra

Eckersly, R. (1981) Biotechnology: It Takes Time to Commercialise, *Sydney Morning Herald*, 19 November, p. 7

Economist, 27 June 1987, pp. 87–93; 19 March 1994
Eder P. (1994) 'Privacy on Parade', *The Futurist*, 28, pp. 38–42
Elmer-Dewitt, P. (1994) The Genetic Revolution, *Time*, 9, 3, pp. 20–7
Ennor, A. (Secretary, Department of Science) (1977) Correspondence to Dr C. Evans (Acting Director-General of Health, Department of Health) 24 June
Enright, J. (ed.) (1983) *The Oxford Book of Death*, Oxford University Press, Oxford
Equal Employment Opportunity Commission (1995) *Compliance Manual*, vol. 2, Section 902, Order 915.002, pp. 902–45; cited in Rothenberg, K. *et al.* (1997) 'Genetic Information and the Workplace: Legislative Approaches and Policy Challenges', *Science*, 275, pp. 1755–7
European Federation of Biotechnology (EFB) (1994) Biotechnology in Food and Drinks, http://www.kluyver.stm.tudelft.nl/tjpb/eng2.htm
Evans, C. (Acting Director-General of Health, Department of Health) (1977) Correspondence to A. Ennor, Secretary, Department of Science, 1 July
Ewing, T. (1994) Patent on Genes Challenged, *The Age*, 9 December, p. 3
Farrands, J. (Secretary Department of Science and the Environment) (1979) Summary of Meeting on Recombinant-DNA (10 August), attached paper to letter dated 21 September 1979 from J. Waterman, Executive Assistant, Dept. of Science and the Environment, to Dr Evans, Australian Academy of Science
Fitt, G. (1994) Transgenic Cotton: Its Place in Integrated Pest Management, Paper Presented at the Sixth Australian Cotton Conference, Gold Coast, August
Fitt, G., Mares, C. and Llewellyn, D. (1994) 'Field Evaluation and Potential Ecological Impact of Transgenic Cotton (*Gossypium hirsutum*) in Australia', *Biocontrol Science and Technology*, 4, pp. 535–48
Forrester, N. (1994) 'Resistance Management Options for Conventional *Bacillus thuringiensis* and Transgenic Plants in Australian Summer Field Crops', *Biocontrol Science and Technology*, 4, pp. 549–53
——(1997) Principal Research Scientist, NSW Agriculture and Australian Cotton Research Institute, pers. comm., January–June
Forrester, N. and Constable, G. (1997), respectively, Principal Research Scientist, NSW Agriculture and Australian Cotton Research Institute; Director, Co-operative Research Centre for Sustainable Cotton, pers comm., 26 June
Foster, J. and Ghonim, S. (1995) 'Prospects for the Commercial Use of Transgenic Plants: Attitudes of Crop Protection Professionals', *Australian Agribusiness Review*, 3, 2, December, pp. 73–94
Fourmile, H. (1995) 'Intellectual Property Rights Need Recognition', *Land Rights Queensland*, October–November, p. 7
——(1996) 'Protecting Indigenous Property Rights in Biodiversity', *Current Affairs Bulletin*, February–March, pp. 36–41
——(1997), pers. comm., 12 August
Fowler, R. (1991) *Language in the News: Discourse and Ideology in the Press*, Routledge, London
Friedman, S. (1986) A Case of Benign Neglect: Coverage of Three Mile Island Before the Accident, in Friedman S., Dunwoody, S. and Rogers, C. (eds) *Scientists and Journalists: Reporting Science as News*, Macmillan, New York
Fuller, K. (DST) (1983) Internal Memo to the Minister, 25 May

Gaita, R. (1997) 'Genocide and Pedantry', *Quadrant*, 338 (July–August), pp. 41–5

Gans, L. (1997) Judged by Your Genes, *The Courier-Mail*, 8 August, p. 18

Geisler, C. and Lyson, T. (1993) 'Bio-technology: Rural Policy Implications of Bovine Growth Hormone Adoption in the USA', in Harper, S. (ed.) *The Greening of Rural Policy: International Perspectives*, Belhaven, London

Gene Exchange (1992) Too Much BT?, 3, 1, p. 6

——(1997) FlavrSavr Tomato—Squashed, Fall, p. 12

Gene File, Genetic Engineering Action Update (1994) The Australian Genethics Network, Australian Conservation Foundation, Melbourne

Gene Technology Information Unit (GTIU) (1995) Gene Technology at Work, *Today's Technology*, 3

Genetic Resources Action International (GRAIN) (1996) The Biotech Battle Over the Golden Crop, *Seedling*, 13, 3, pp. 23–32

George, S. (1976) *How the Other Half Dies*, Penguin, England

Giddens, A. (1991) *Modernity and Self-Identity*, Polity Press, Cambridge

Gilchrist, G. (1995) Gene Therapy May be Used to Treat Brain Disorders, *Sydney Morning Herald*, 28 October, p. 3

Glanzig, A.(1996) Magnificent Seven is a Natural Act, *Australian*, 5 June, p. 18

Glover, J., 1984. *What Sort of People Should There Be?* Penguin, Harmondsworth

Godard, B. and Verhoef, M. (1996) 'DNA Banking: Current and Ideal Practices', in Knoppers, B., Caulfield, T. and Kinsella, T. (eds) *Legal Rights and Human Genetic Material*, Emond Montgomery Publications, Toronto

Goleby, R. (Assistant Secretary Policy Division Department of Science) (1978) Internal Correspondence to the Secretary, Dept of Science (77/537)

Goodell, R. (1986) 'How to Kill a Controversy: The Case of Recombinant DNA', in Friedman, S., Dunwoody, S. and Rogers, C. (eds) *Scientists and Journalists: Reporting Science as News*, Macmillan, New York

Goodman, D. and Redclift, M. (1991) *Refashioning Nature: Food, Ecology and Culture*, Routledge, London

Gottweis, H. (1995) 'German Politics of Genetic engineering and Its Deconstruction', *Social Studies of Science* 25, pp. 195–235

Gould, F., Martinez-Ramirez, A., Anderson, A., Ferre, J., Silva, F. and Moar, W. (1992) 'Broad-spectrum Resistance to *Bacillus thuringiensis* Toxins in *Heliothis virescens*', *Proceedings of the National Academy of Sciences, USA* 89, pp. 7989–90

Gray, I., Stehlik, D. and Lawrence, G. (1997) Economic Management, Social Contradiction and Climatic Disaster: Drought Policy in the 1990s, Paper Presented at the Rural Australia 2000 Conference, Charles Sturt University, Wagga Wagga, 2–4 July

Green, R. (DST) (1984) Interdepartmental correspondence, 19 September

Greens in the European Parliament n.d. Opposition to European Patent No. 0112149 (Howard Florey Institute of Experimental Physiology and Medicine, University of Melbourne, Australia), Submission in the European Patent Office

Griffiths, A., Miller, J., Suzuki, D., Lewontin, R. and Gelbart, W. (1996) *An Introduction to Genetic Analysis*, 6th edn, WH Freeman and Company, New York

Grifo, F., Newman, D., Fairfield, A., Bhattacharya, B. and Gruppenhoff, J.

(1997) 'The Origins of Prescription Drugs', in Grifo, F. and Rosenthal, J. (eds) *Biodiversity and Human Health*, Island Press, Washington, DC

Hall, J., Lee M., Newman B., Morrow J., Anderson L., Huey B. and King M-C. (1990) 'Linkage of Early-onset Familial Breast Cancer to Chromosome 17q21', *Science*, 250, pp. 1684–89

Hallman, W. and Metcalfe, J. (1993) *Public Perceptions of Agricultural Biotechnology: A Survey of New Jersey Residents*, State University of New Jersey, New Brunswick

Hannigan, J. (1995) *Environmental Sociology: A Social Constructivist Perspective*, Routledge, London

Hansard (1977) Senate, 23 February, p. 283

Harper, S (ed.) (1993) *The Greening of Rural Policy*, Belhaven Press, London

Haynes, R. (1991) *High Tech, High Co$t*, Pan Macmillan Australia, Sydney

——(1994) *From Faust to Strangelove: Representations of the Scientist in Western Literature*, John Hopkins University Press, Baltimore

Hazen, R. and Trefil, J. (1990) *Science Matters. Achieving Scientific Literacy*, Doubleday, New York

Healy, M. (1991) *Strategic Technologies for Maximising the Competitiveness of Australia's Agriculture-Based Exports*, IP/2/91, Bureau of Rural Resources, Canberra

Henningham, J. (1995) Who are Australia's Science Journalists? *Search*, 26, 3, pp. 89–94

Herman, E. and Chomsky, N. (1994) *Manufacturing Consent: The Political Economy of the Mass Media*, Vintage, London

Hindmarsh, R. (1992a) 'CSIRO's Genetic Engineering Exhibition: Public Acceptance or Public Awareness?', *Search*, 23, 7, pp. 212–3

——(1992b) 'Agricultural Biotechnologies: Ecosocial Concerns for a Sustainable Agriculture', in Lawrence, G., Vanclay, F. and Furze, B. (eds) *Agriculture, Environment and Society: Contemporary Issues for Australia*, Macmillan, Melbourne

——(1994) Power Relations, Social-Ecocentrism, and Genetic Engineering: Agrobiotechnology in the Australian Context (unpub. doctoral thesis Griffith University)

——(1996) 'Bio-Policy Translation in the Public Terrain', in Lawrence, G., Lyons, K. and Momtaz, S. (eds) *Social Change in Rural Australia: Perspectives from the Social Sciences*, Central Queensland University, Rockhampton

——(1998) 'Globalisation and Gene-Biotechnology: From the Centre to the Semi-Periphery', in Burch, D., Rickson, R. and Lawrence, G. (eds) *Australasian Food and Farming in a Globalised Economy: Recent Developments and Future Prospects*, Monash University, Melbourne

Hindmarsh, R., Burch, D. and Hulsman, K. (1991) 'Agrobiotechnology in Australia: Issues of Control, Collaboration and Sustainability', *Prometheus* 9, 2, pp. 221–48

Hindmarsh, R. and Hulsman, K. (1992) 'Gene Technology: the Threat or the Glory?', *New Scientist,* (Australian supplement) 25 April, p. 4

Hindmarsh, R., Lawrence, G. and Norton, J. (1995) 'Manipulating Genes or Public Opinion?' *Search*, 26, pp. 117–21

Hoad, B. (1977) 'The Dangers of Life from a Test-tube: The New Frontier in Australian Science', *Bulletin*, 17 September, pp. 38–44

Hoban, T. (1989) 'Anticipating Public Response to Biotechnology', *The Rural Sociologist* 9, 3, pp. 20–4

Hoban, T., Woodrum, E. and Czaja, R. (1992) 'Public Opinion to Genetic Engineering', *Rural Sociology* 57, 4, pp. 476–93

Hobbelink, H. (1991) *Biotechnology and the Future of World Agriculture*, Zed Books, London

Holman, H. and Dutton, D. (1978) 'A Case for Public Participation in Science Policy Formation and Practice', *Southern California Law Review* 51, pp. 1505–34

Holub, M. (1990) *The Dimension of the Present Moment: And Other Essays* (ed. and trans. David Young), Faber and Faber, London

——(1992) 'Jumping to Conclusions', *Island*, 53, pp. 16–21

——(1993) 'This Long Disease', *Island*, 54, pp. 4–6

Horin, A. (1976) Genetics—the Myth Becomes Reality, *The National Times*, 11–16 October, pp. 16–17

Hornig Priest, S. (1995) 'Information Equity, Public Understanding of Science, and the Biotechnology Debate', *Journal of Communication*, 45, 1, pp. 39–54

Howarth, F. (1984) Biotechnology: Revolutionary or Evolutionary?, unpub. MSc thesis, NSW University

Hoy, A. (1996a) Groups Round Up Support for "Gene Bean" Supermarket Warning, *Sydney Morning Herald*, 16 December, p. 3

——(1996b) Awassi Project Aims to Put Home-grown Roquefort on Menu, *Sydney Morning Herald*, 16 April, p. 9

——(1996c) Handless Robot's Green Thumbs-up for the Forests, *Sydney Morning Herald*, 5 November, p. 4

Hubbard, R. (1988) 'Eugenics: New Tools, Old Ideas', in Hoffmann Baruch, E., D'Adamo, A. and Seager, T. (eds) *Embryos, Ethics, and Women's Rights: Exploring the New Reproductive Technology*, The Haworth Press, New York

——(1995) 'Human Nature', in Shiva, V. and Moser, I. (eds) *Biopolitics*, Zed Books, London

——and Wald, E. (1997) *Exploding the Gene Myth*, Beacon Press, Boston

Hudson, K., Rothenberg, K., Andrews L., Ellis Kahn, M. and Collins, F. (1995) 'Genetic Discrimination and Health Insurance: An Urgent Need for Reform', *Science*, 270, pp. 391–3

Hughes, S. (1986) A Good Test but Where Can You Take It?, *Guardian*, 7 July

Hutton, P. (1997) USEPA Biopesticide and Pollution Prevention Division pers comm., 17 January

Huxley, A. (1977) *Brave New World*, Grafton Books, London

Hynes, H. (1989) 'Biotechnology in Agriculture: An Analysis of Selected Technologies and Policy in the United States', *Reproductive and Genetic Engineering*, 2, 1, pp. 39–49

INRA (1991) Opinions of Europeans on Biotechnology in 1991, Commission of the European Communities, Brussels

International Agricultural Development (1996) Organic Focus: Expanding Supply and Demand, 16, 6, p. 23

Invetech (1989) Opportunities in Biotechnology for Australian Industry: an Overview, Invetech Operations, Victoria

Irwin, A. (1997) Call to Review Green Gene Rulebook, *Daily Telegraph*, 10 March, London

Jackson, W. (1984) 'A New Threat to Agriculture and a Window of Opportunity', *Ecologist*, 14, pp. 119–24

James, C. and Krattiger, A. (1996) Global Review of the Field Testing and Commercialization of Transgenic Plants: 1986 to 1995, International Service for the Acquisition of Agri-biotech Applications, New York

Jank, B. (1995) 'Biotechnology in European Society', *Trends in Biotechnology*, 13, pp. 42–4

Jennings, P. (1994) 'Biotechnology and the Biological Indiustries in the 21st Century', in Eckersley, R. and Jeans, K. (eds) *Challenge to Change: Australia in 2020*, CSIRO Publications, Australia

Joseph, R. and Johnston, R. (1985) 'Market Failure and Government Support for Science and Technology: Economic Theory Versus Political Practice', *Prometheus* 3, 1, pp. 138–55

Joyce, C. (1990) 'US Approves Trials with Gene Therapy', *New Scientist*, 11 August, p. 7

Judson, H. (1979) *The Eighth Day of Creation: Makers of the Revolution in Biology*, Simon and Schuster, New York

Katz Rothman, B. (1986) *The Tentative Pregnancy*, Penguin, New York

Kauffman, S. (1993) *Origins of Order; Self-organization and Selection in Evolution*, Oxford University Press, New York

——(1996) Investigations, Technical Report #96-08-072, Santa Fe Institute, Santa Fe, New Mexico, USA

Kawar, A. (1989) 'Issue Definition, Democratic Participation and Genetic Engineering', *Policy Studies Journal*, 17, 4, pp. 719–44

Kelley, J. (1995) Public Perceptions of Genetic Engineering: Australia, 1994, Final Report to the Department of Industry, Science and Technology, Institute of Adnaced Studies, Australian National University, Canberra

Kidwell, M. and Lisch, D. (1997) 'Transposable Elements as Sources of Variation in Animals and Plants', *Proceedings of the National Academy of Sciences USA* 94, pp. 7704–11

Kloppenburg, J. (1988) *First the Seed: The Political Economy of Plant Biotechnology*, Cambridge University Press, Cambridge

——(1991) 'Alternative Agriculture and the New Biotechnologies', *Science as Culture*, 13, pp. 482–505

Kloppenburg, J. and Burrows, B. (1996) 'Biotechnology to the Rescue? Twelve Reasons Why Biotechnology is Incompatible with Sustainable Agriculture', *The Ecologist*, 26, 2, March–April, pp. 61–7

Knoppers, B., Caulfield, T. and Kinsella, T. (eds) (1996) *Legal Rights and Human Genetic Material*, Emond Montgomery, Toronto

Kretzschmar, H, Stowring, L, Westaway, D., Stubblebine, W., Prusiner, S. and Dearmond, S. (1986) 'Molecular Cloning of a Human Prion Protein cDNA', *DNA*, 5, 4, pp. 315–24

Krimsky, S. (1982) *Genetic Alchemy: The Social History of the Recombinant-DNA Controversy*, MIT Press, Cambridge, Massachusetts

——(1991) *Biotechnics and Society: The Rise of Industrial Genetics*, Praeger, New York

——and Wrubel, R.(1996) *Agricultural Biotechnology and the Environment: Science, Policy and Social Issues*, University of Illinois, Urbana

Kutujkian, G. (Secretary to UNESCO's International Bioethics Committee) (1996) Oral response to a question at the launch of the Draft Declaration at the United Nations, Paris

Landline (1997) Australian Broadcasting Corporation, 4 May

Lasker, J. and Borg, S. (1987) *In Search of Parenthood: Coping With Infertility and High-Tech Conception*, Beacon Press, Boston

Latour, B. (1987) *Science in Action: How to Follow Scientists and Engineers through Society*, Open University Press, Cambridge

Law Reform Commission of Victoria (1988) Genetic Manipulation, Discussion Paper No. 11, Law Reform Commission of Victoria, Melbourne

Lawrence, G. (1987) *Capitalism and the Countryside: The Rural Crisis in Australia*, Pluto, Sydney

——(1995) Futures for Rural Australia: From Agricultural Productivism to Community Sustainability, Inaugural Address, Central Queensland University, RSERC, Rockhampton

Larence, G. and Norton, J. (1994) Industry Involvement in Australian Biotechnology: the Views of Scientists, *Australasian Biotechnology*, 4, 6, pp. 362–8

Lawrence, G. and Vanclay, F. (1992) 'Agricultural Production and Environmental Degradation in the Murray-Darling Basin', in Lawrence, G., Vanclay, F. and Furze, B. (eds) *Agriculture, Environment and Society: Contemporary Issues for Australia*, Macmillan, Melbourne

Lawrence, G., Vanclay, F. and Furze, B. (eds) (1992) *Agriculture, Environment and Society: Contemporary Issues for Australia*, Macmillan, Melbourne

Lawrence, G., Lyons, K. and Momtaz, S. (eds) (1996) *Social Change in Rural Australia*, RSERC, Central Queensland University, Rockhampton

Lederberg, J. (1991) 'Pandemic as a Natural Evolutionary Phenomenon', in Mack, A. (ed.) *In Time of Plague. The History and Social Consequences of Lethal Epidemic Disease*, New York University Press, New York

Lee, M. (1992a) Genetic Manipulation: the Threat or the Glory?, House of Representatives. Standing Committee on Industry, Science and Technology, Australian Government Publishing Service, Canberra

——(1992b) Media Release: Inquiry into Genetically Modified Organisms, House of Representatives Standing Committee on Industry, Science and Technology, 26 March

Lemkow, L. (1993) Public Attitudes to Genetic Engineering: Some European Perspectives, European Foundation for the Improvement of Living and Working Conditions, Luxembourg

Lenoir, N. (President of the International Bioethics Committee of the United Nations Educational, Scientific and Cultural Organisation) (1996) Presentation of the Preliminary Draft of a Universal Declaration on the Human Genome and Human Rights, 11 September (limited private distribution)

Lester, I. (1994) *Australia's Food and Nutrition*, Australian Government Publishing Service, Canberra

Levidow, L. (1996) 'Simulating Mother Nature, Industrialising Agriculture', in Robertson, G., Marsh, M., Tickner, L., Bird, J., Curtis, B. and Putnam, T. (eds) *FutureNatural: Nature, Science, Culture*, Routledge, London

Lewin, B. (1994) *Genes V*, Oxford University Press, Oxford
Lifton, R. (1987) *The Nazi Doctors*, Macmillan, London
Llewellyn, D (1997) (CSIRO Division of Plant Industry), pers. comm., 28 January; 6 June
Llewellyn, M. (1995) How the Genes Revolution Poses Human Dilemmas, *The Northern Herald*, 21 September, p. 6
Lloyd, G. (1997) Farewell To Privacy, *The Courier Mail*, 12 April, p. 26
Loge, P. (1991) Language is a Virus: Discourse and the Politics and Public Policy of Biotechnology, Paper prepared for the 12th International Meeting of the Society of Environmental Toxicology and Chemistry, Seattle, US, 6 November
Love, R. (1993) 'The Public Relations of Science, the Flying Pink Pig and the Jet-propelled Cane-toad', *Chain Reaction*, 68, pp. 21–2
Lövel, G. (1997) 'Global Change Through Invasion', *Nature*, 388, pp. 627–8
McAuliffe, S. and McAuliffe, K. (1981) *Life for Sale*, Coward, McCann & Geoghegan, New York
McEwen, J. and Reilly, P. (1994) 'Stored Guthrie Cards as DNA "Banks"', *American Journal of Human Genetics* 55, pp. 196–200
McLaren, A. (1987) 'Can We Diagnose Genetic Disease in Pre-embryos?', *New Scientist*, l0 December, p. 42
McLean, G., Waterhouse, P., Evans, G. and Gibbs, M. (1997) *Commercialisation of Transgenic Crops: Risk, Benefit and Trade Considerations*, (Department of Primary Industries and Energy/Bureau of Resource Sciences) Australian Government Publishing Service, Canberra
McNeill, P. (1993) *Ethics and Politics of Human Experimentation*, Cambridge University Press, Melbourne
McNulty, J. (1997) Australian Farmers Fill the Demand for Conventional Crops, 17 August, source -email address-gentech@tribe.ping.de
Maddox, J. (1993) 'Wilful Public Misunderstanding of Genetics', *Nature*, 364, p. 281
Malcolm, B., Sale, P. and Egan, A. (1996) *Agriculture in Australia: an Introduction*, Oxford University Press, Melbourne
Malkin, D., Li F., Strong L., Fraumeni J. Jr., Nelson C., Kim D., Kassel J., Gryka M., Bischoff F., Tainsky M. and Friend S. (1990) 'Germ line *p53* Mutations in a Familial Syndrome of Breast Cancer, Sarcomas, and Other Neoplasms', *Science*, 250, pp. 1233–8
Marsden, T. (1992) 'Exploring a Rural Sociology for the Fordist Tradition', *Sociologia Ruralis*, XXXII, 2–3, pp. 209–30
Martin, E. (ed.) (1990) *A Concise Dictionary of Biology*, Oxford University Press, London
Marx, J. (1993) 'Cell Death Studies Yield Cancer Clues', *Science*, 259, pp. 760–l
Masood, E. (1996) 'Gene Tests: Who Benefits From Risk?', *Nature*, 379, pp. 389–92
Mayeno, A. and Gleich, G. (1994) 'Eosinophilia-myalgia syndrome and tryptothan production: a cautionary tale', *Trends in Biotechnology* 12, pp. 346–52
Mazur, A. (1981) 'Media Coverage and Public Opinion on Scientific Controversies', *Journal of Communication*, 31, pp. 106–15
Mensah, R. (1996) (IPM researcher, NSW Agriculture) pers. comm.

Millis, N. (Chair, r-DNA Monitoring Committee) 1986 Correspondence to Prof D. Shanks, Vice Chancellor, Adelaide University, 9 July
Millis, N. (Chairman RDMC) (1984) Correspondence to Barry Jones (Minister for Science and Technology, 17 December
Mills, R. (1990) Decision Time for Genetic Engineering, *Australian Rural Times*, 21–27 June, p. 6
Mitcham, C. (1997) 'Justifying Public Participation In Technical Decision Making', *IEEE Technology and Society Magazine* 16, 1, pp. 40–6
Monsanto (1995) *Introducing Ingard: Some Basic Facts and Perspectives*, Monsanto Australia, Melbourne
——(1995) Roundup Ready Gene Agreement 1996, Monsanto Agreement for Roundup Ready Soybeans, Monsanto
——(n.d.) Results of Bollgard—Research Regarding Bollworm Infestations, Monsanto
——(1997) Effective Weed Control
Morton, O. (1997) 'First Dolly, Now Headless Tadpoles', *Science* 278, 5339, p. 798
Moser, I. (1995) 'Critical Communities and Discourses on Modern Biotechnology', in Shiva, V. and Moser, I. (eds) *Biopolitics*, Zed Books, London
Mulkay, M. (1993) 'Rhetorics of Hope and Fear in the Great Embryo Debate', *Social Studies of Science*, 23, pp. 721–42
Mussared, D. (1994) Yes, We'll Eat those Tomatoes, *Canberra Times*, 18 October, p. 12
Myers, G. (1990) 'The Double Helix as Icon', *Science as Culture*, 9, p. 49
National Action Plan on Breast Cancer (NAPBC) and the NIH–DOE Working Group on the Ethical, Legal, and Social Implications of Human Genome Research. (1995) Genetic Discrimination and Health Insurance: A Case Study on Breast Cancer', Bethesda, MD, 11 July 1995, Workshop (cited in Hudson, K. *et al.* (1995) 'Genetic Discrimination and Health Insurance: An Urgent Need for Reform', *Science*, 270, pp. 391–3)
National Farmers' Federation (1993) New Horizons: A Strategy for Australia's Agrifood Industries, Canberra
——(1996) 1996 *Annual Report*, Canberra
National Health and Medical Research Council (NHMRC) (1991) *Guidelines for the Use of Genetic Registers in Medical Research*, Australian Government Printer, Canberra
——(1995) *Aspects of Privacy in Medical Research*, Australian Government Printer, Canberra
National Occupational Health and Safety Commission (1994) *Control of Workplace Hazardous Substances: National Model Regulations for the Control of Workplace Hazardous Substances*, Australian Government Publishing Service, Canberra
National Registration Authority (NRA) (1996a) Public Release Summary: INGARD Gene by Monsanto, National Registration Authority for Agricultural and Veterinary Chemicals, Canberra, 7 May
——(1996b) NRA Community Brief, National Registration Authority for Agricultural and Veterinary Chemicals, Canberra, 6 August
Nature (1992) 'Genetics and the Public Interest', 356, 2 April, pp. 365–6

——(1997) 'Council of Europe Urged to Ban Human Cloning', 389, 6652, p. 656

Nature Biotechnology (1996) 'Soya, Supermarkets, Sense and Segregation', 14 December, p. 1627

Neale, G. (1997) 'Frankenstein Food' Faces Supermarket Ban, *Sunday Telegraph*, 26 January, London

Nelkin, D. (1987) *Selling Science: How the Press Covers Science and Technology*, W.H. Freeman and Company, New York

——(1996) 'Genetics, God and Sacred DNA', *Society*, 33, 4, pp. 22–5

Nelkin, D and Lindee, M. (1995) *'The DNA Mystique: The Gene as a Cultural Icon'*, W.H. Freeman, New York

New York Times (1977) 24 July

Norton, J. and Lawrence, G. (1996) 'Consumer Attitudes to Genetically Engineered Food Products: Focus Group Research in Rockhampton, Queensland', in Lawrence, G., Lyons, K. and Momtaz, S. (eds) *Social Change in Rural Australia*, Rural Social and Economic Research Centre, (RSERC), Central Queensland University, Rockhampton

Norton, J., Lawrence, G. and Wood, G. (forthcoming 1998) 'The Australian Public's Perception of Genetically-Engineered Foods', *Australasian Biotechnology*, in press

Nurse, P. (1997) 'The Ends of Understanding', *Nature*, 387, p. 657

O'Connor K. (1996) *The Privacy Implications of Genetic Testing, Information Paper Number Five*, Commonwealth of Australia, Canberra

OECD (Group of National Experts on Safety in Biotechnology) (1992) Public Information/Public Education in Biotechnology: Results of an OECD Survey, OECD, DSTI/STP/BS(92)7, Paris

Office of Technology Assessment (OTA). (1987) New Developments in Biotechnology: Public Perceptions of Biotechnology, US Congress Office of Technology Assessment, Washington

Official Journal of Patents (1976) 'Trade Marks and Designs', 21 October, p. 3915

O'Neill, G. (1990a) Uproar over mutant meat, *The Age*, 28 April, p. 1

——(1990b) The Issue is the Right to Know, *The Age*, 2 May, p. 13

——(1991) 'Human DNA: Cracking the Code', *21C*, Summer 90–91, pp. 39–43

Osfield, S. (1997) 'Foods of the Future', *Australian Good Taste Magazine*, February, 1997, pp. 52–3

Passey, D. (1996) Beef's Image Takes a Hiding, *Sydney Morning Herald*, 20 January, p. 39

Patterson, P. and Wilkins, L. (1991) *Media Ethics: Issues and Cases*, Wm. C. Brown Publishers, Dubuque

——(1994) 'Genetic Engineering of Crop Plants Will Enhance the Quality and Diversity of Foods', *Food Australia*, 46, pp. 379–81

Peacock, J. (CSIRO Division of Plant Industry) (1990) Internal Correspondence to J. Stocker (Chief Executive CSIRO), 7 June

——(1995) Gene Technology and Our Future Lifestyle, Australian Foundation for Science Lecture presented at ANZAAS '95 Congress, Newcastle, August

Penman, D. (1997) 'Gag' on Gene-altered Food, *The Guardian Weekly* , 17 August, p. 10

Petersen, A. (1996) 'Risk and the Regulated Self: The Discourse of Health

Promotion as Politics of Uncertainty', *Australian and New Zealand Journal of Sociology*, 32, 1, pp. 44–57
Phelps, R. (1997) Business is Eating Away at our Food Supply, *Canberra Times*, 16 October, p. 11
Playne, M. (1998) 'Public Perceptions of Genetic Engineering', *Australasian Biotechnology* 8, 1, pp. 39–42
Plein, L. (1991) 'Popularizing Biotechnology: The Influence of Issue Definition', *Science, Technology, and Human Values*, 16, 4, pp. 474–90
Pocock, I. (1997) GE-Irish Update, 15 August, source-email address: gentech@tribe.ping.de
Pokorski, R. (1995) 'Genetic Information and Life Insurance', *Nature*, 376, pp. 13–4
Prakash, C. (1997) A Commentary on the Proceedings of the National Academy of Science Paper Concerning Bt Cotton Adaptation by Insects, from *ISB NewsReport*, July (http://www.nbiap.vt.edu)
Priest, S. (1995) 'Information Equity, Public Understanding of Science and the Biotechnology Debate', *Journal of Communication*, 45, 1, pp. 39–54
Prusiner, S. (1997) 'Prion Diseases and the BSE Crisis', *Science*, 278, pp. 245–51
Pure Food Campaign (1997a) Kagome Suspends Plan to Market Foods From Bio-tomatoes, source-email address-gentech@tribe.ping.de
——(1997b) Internet Alert, Threatened Corporate Take-over of Organic Standards in the USA, Pure Food Campaign
Pyke, B. (1997) (Entomologist, Cotton Research and Development Corporation) pers comm., January–July
Quade, V. (1993) 'Protecting the Essence of Being', *Human Rights*, 20, pp. 14–17
Quantum (1996) Australian Broadcasting Corporation, 15 August
Raymond, J. (1993) *Women As Wombs: Reproductive Technologies and the Battle Over Women's Freedom*, Harper, San Francisco
Recombinant-DNA Monitoring Committee (RDMC) (Department of Science of Technology) (1982) *Recombinant DNA Monitoring Committee Report for the Period October 1981 to October 1982*, Australian Government Publishing Service, Canberra
Reeves, T. (Chairman, Advisory Committee, CSIRO Division of Plant Industry) (1990) Internal Correspondence to J. Stocker (Chief Executive CSIRO) 23 May
Rifkin, J. (1984) *Algeny*, Penguin, Harmondsworth
Ripe, C. (1996) Designer Genes, No Label, *Australian*, 13 December, p. 17
Rissler, J. (1997) 'Bt Cotton-Another Magic Bullet?', *Global Pesticide Campaigner*, 7, 1
Rissler, J. and Mellon, M. (1990) 'Public Access to Biotechnology Applications', *Natural Resources and Environment*, pp. 4, 3, pp. 29–31, 54–8
Roberts, L. (1990) 'L-tryptophan Puzzle Takes New Twist', *Science*, 249, 31 August, p. 988
Rose, S. and Rose, H. (1976) 'The Politics of Neurobiology: Biologism in the Service of the State', in Rose, H. and Rose, S. (eds) *The Political Economy of Science*, MacMillan, UK
Rothblatt, M. (1997) *Unzipped Genes: Taking Charge of Baby-Making in the New Millennium*, Temple University Press, Philadelphia

Rothenberg, K., Fuller, B., Rothstein, M., Duster, T., Ellis Kahn, M., Cunningham, R., Fine, B., Hudson, K., King, M-C., Murphy, P., Swergold, G. and Collins, F. (1997) 'Genetic Information and The Workplace: Legislative Approaches and Policy Challenges', *Science*, 275, pp. 1755–7

Roush, R. (1997) Associate Professor, Department of Crop Protection, Waite Institute, University of Adelaide, pers. comm., January–July

——(1997b) Bogus GENETIX Press Release?, rroush@waite.adelaide.edu.au (polant@saclink.csus.edu), 10 October

Rowe, P. (1996) Lorimer Dods Professor and Director's Report, The 1996 Children's Medical Research Institute Annual Report, The Children's Medical Research Institute, Sydney

——(1997) pers. comm., July 1997

Rowell, A. (1996) *Green Backlash: Global Subversion of the Environment Movement*, Routledge, London

Rowland, R. (1992) *Living Laboratories: Women and Reproductive Technologies*, Pan Macmillan Australia, (distributed by Spinifex Press, Melbourne)

Rural Advancement Foundation International (RAFI) (1989) Biotechnology Industry Consolidation, *RAFI Communique*, November, pp. 1–12

——(1994) Declaring the Benefits, *RAFI Occasional Paper*, 1(3), pp. 1–14

——(1996a) The Life Industry, *RAFI Communique*, September, pp. 1–8

——(1996b) Pharmaceutical Companies Bid for Northern Botanical Garden Collections in Attempt to Avoid the Biodiversity Convention, *RAFI Communique*, July–August, p. 10

——(1997) Bioserfdom: Technology, Intellectual Property and the Erosion of Farmers' Rights in the Industrialized World, *RAFI Communique*, March–April, pp. 1–10

Saltus, R. (1986) Biotech Firms Compete in Genetic Diagnosis, *Science*, 234, p. 1319

Saxton, M. (1988) 'Prenatal Screening and Discriminatory Attitudes About Disability', in Hoffmann Baruch, E., D'Adamo, A. and Seager, T. (eds), *Embryos, Ethics, and Women's Rights: Exploring the New Reproductive Technology*, The Haworth Press, New York

Schattschneider, E. (1960) *The Semi-Sovereign People*, Holt, Rinehart and Winston, New York

Schmeck, H. (1986) Crystal Ball is Ethically Dark, *The Age*, 3 November

Schmidt, K. (1997) 'It Was My Genes Guv', *New Scientist*, 2107, pp. 46–50

Scott-Kemmis, D., Darling, T., Johnston, R., Collyer, F. and Cliff, C. (1990) *Strategic Alliances in the Internationalisation of Australian Industry*, Australian Government Publishing Service, Canberra

Scott-Kemmis, D. and Darling, T. (1991) Biotechnology and Processed Food, *Australasian Biotechnology*, 1, 1, pp. 32–5

Shand, H. (1997) Human Nature: Agricultural Biodiversity and Farm-based Food Security, United Nations Food and Agriculture Organization, Rome

Shapin, S. (1988) 'Following Scientists Around', *Social Studies of Science* (Essay Review), 18, pp. 533–50

Shapiro, J. (1997) Genome Organization, Natural Genetic Engineering and Adaptive Mutation, *Trends in Genetics*, 13, 3, pp. 98–104

Sharp, S. (1992) 'Cottoning On', *Simply Living*, 7, 2, pp. 17–18

Shiva, V. (1997) 'Dr Vandana Shiva Responds', *The Ecologist* (Letter Forum), 27, 5, pp. 211–2

Shiva, V. and Moser, I. (1995) *Biopolitics*, Zed Books, London

Sleigh, M. (1988) 'The Future of Commercial Biotechnology', *Australian Journal of Biotechnology*, 2, 1, pp. 26–7

Smith, D. (1995) One Man's Battle in Melanoma War, *Sydney Morning Herald*, 9 December, p. 27

Smith, N. (1996) 'The Production of Nature', in Robertson, G., Marsh, M., Tickner, L., Bird, J., Curtis, B. and Putnam, T. (eds) *FutureNatural: Nature, Science, Culture*, Routledge, London

Solé, R., Manrubia, S., Benton, M. and Bak, P. (1997) 'Self-similarity of Extinction Statistics in the Fossil Record', *Nature*, 388, pp. 764–7

Sparks, S., Shepherd, R. and Frewer, L. (1994) 'Gene Technology, Food Production and Public Opinion: A UK Study', *Agriculture and Human Values*, 19, 1, pp. 19–28

Stanworth, M. (1987) 'The Deconstruction of Motherhood', in Stanworth, M. (ed.) *Reproductive Technologies: Gender, Motherhood and Medicine*, Polity Press, Cambridge

Steinbrecher, R. (1996) 'From Green Revolution to Gene Revolution: The Environmental Risks of Genetically Engineered Crops', *The Ecologist*, *26*, 6, pp. 273–81

Stocker, J. (1990) (Chief Executive CSIRO) Internal Correspondence to T. Reeves (Chairman, Advisory Committee, CSIRO Division of Plant Industry), 4 June

——(1997) *Priority Matters*, Department of Industry, Science and Tourism, Corporate Communications Section, Canberra

Stott Despoja, N. (1997) Mapping Out the Future of Genes, *Canberra Times*, 18 November, pp. 43–5

Suzuki, D., and Knudston, P. (1988) *Genethics: The Ethics of Engineering Life*, Allen & Unwin, Sydney, Australia, originally published by Stoddart Publishing Co. Ltd, Toronto, Canada

Suzuki, D. and Levine, J. (1994) *Cracking the Code: Redesigning the Living World*, Allen & Unwin, Sydney, Australia

Sweet, M. (1995) New Study Shows Families with Heart Attack Gene More at Risk, *Sydney Morning Herald*, 16 March, p. 2

——(1997) Old Drugs to Cure Modern-day Ills, *Sydney Morning Herald*, 25 February

Swinbanks, D. and Anderson, C. (1992) Search for Contaminant in EMS Outbreak Goes Slowly, *Nature*, 358, 9 July, p. 96

Sydney Morning Herald 1 April 1993; 19 June 1995; 21 July 1995; 16 December 1995; 20 October 1997, p. 11

Tait, J. (1990) 'Environmental Risks and the Regulation of Biotechnology', in Lowe, P., Marsden, T. and Whatmore, S. (eds) *Technological Change and the Rural Environment*, Fulton, London

Third World Network (1995) *The Need for Greater Regulation and Control of Genetic Engineering: A Statement by Scientists Concerned about Current Trends in the New Biotechnology*, Third World Network, Penang, Malaysia

Tiffen, R. (1989) *News and Power*, Allen & Unwin, Sydney

Time 17 January 1994; 25 April 1994

Troutwine, G. (1991) 'Genetic Engineering of L-tryptophan: Futuristic Disaster', *Trial* July, pp. 20–5

Tudge, C. (1993) *The Engineer in the Garden: Genetics—From the Idea of Heredity to the Creation of Life*, Pilmico, London

Turney, J. (1993) 'Thinking About the Human Genome Project', *Science as Culture*, 19, p. 293

TWN (Third World Network) (1995) The Need for Greater Regulation and Control of Genetic Engineering: A Statement by Scientists Concerned about Current Trends in the New Biotechnology, Penang, TWN, Malaysia

Uhlig, R. (1996) Gene Engineered Foods are Unsafe, Says Scientist, *Daily Telegraph*, 7 September, London

United Nations Educational, Scientific and Cultural Organisation (1996) *Draft Declaration on the Human Genome and Human Rights*, UNESCO, Paris

United Nations (UN) (1994) *1994 Human Rights: A Compilation of International Instruments*, vol. 1 (First Part), Centre for Human Rights, Geneva

University of Melbourne Assembly Report (1979) Report on Genetic Engineering, Assembly Reports

University of Melbourne (1994) The University of Melbourne Research Report 1994

US Patent Office (1997) Patent Number 5672607, Lexis-Nexus Database, 30 September

Vanclay, F. and Lawrence, G. (1995) *The Environmental Imperative: Eco-social Concerns for Australian Agriculture*, CQU Press, Rockhampton

Venter, C. and Cohen, D. (1997) Genetic Code-Breakers, *Weekend Australian*, 19–20 July, p. 28

Verkerk, R. (1997) IPM researcher, Imperial College of Science, Technology and Medicine, Silwood Park, UK, pers. comm., July

Verkerk, R. and Wright D. (1997) 'Field-based Studies on Diamondback Moth Tritrophic System in Cameron Highlands, Malaysia', *International Journal of Pest Management*, 43, pp. 27–33

Verhoef, M., Lewkonia, R. and Kinsella, D. (1996) 'Ethical Implications of Current Practices in Human DNA Banking in Canada', in Knoppers, B., Caulfield, T. and Kinselle, T. (eds) *Legal Rights and Human Genetic Material*, Edmond Montgomery Publication, Toronto

Verma, I. and Somia, N. (1997) Gene Therapy-Promises, Problems and Prospects, *Nature*, 389, 18 September, pp. 239–42

Vines, G. (1986) 'Test-tube Pioneer Fears Rise of Eugenics', *New Scientist*, 9 October, p. 17

Wagner, W., Torgerson, H., Einsiedel, E., Jelsoe, E., Fredickson, H., Lassen, J., Rusanen, T., Boy, D., de Cheveigne, S., Hampel, J., Stathopoulou, A., Allansdotottir, A., Midden, C., Nielsen, T., Przestalski, A., Twardowski, T., Fjaestad, B., Olsson, S., Olofsson, A., Gaskell, G., Durant, J., Bauer, M. and Liakopoulos, M. (1997) 'Europe Ambivalent on Biotechnology', *Nature*, 387, pp. 845–7

Waldrop, M. (1992) *Complexity: The Emerging Science at the Edge of Order and Chaos*, Simon and Schuster, New York

Ward, I. (1995) *Politics of the Media*, MacMillan, Melbourne

Ward, S. (1977) Correspondence to RAFI, 13 June

Webster, G. and Goodwin, B. (1996) *Form and Transformation: Generative and Relational Principles in Biology*, Cambridge University Press, Cambridge

Weinberger, J. (1989) *New Atlantis and The Great Instauration Bacon*, rev. edn, Harlan Davidson, Illinois

Wells, H.G. (1975) *The Island of Dr. Moreau*, Pan, London

Wells, J. (1995) Genetic Testing May Unearth Gold, *Sydney Morning Herald*, 9 October, p. 44

Wheale, P. and McNally, R. (1988) *Genetic Engineering: Catastrophe or Utopia?*, Harvester, Wheatsheaf

——(1990) *The Bio-Revolution: Cornucopia or Pandora's Box*, Pluto Press, London

Wilkie, T. (1995) Genetic Link with Petty Crime Claimed, *Sydney Morning Herald*, 16 February, p. 29

Wilkie, T. (1996) 'Genes "R" Us', in Robertson, G., Marsh, M., Tickner, L., Bird, J., Curtis, B. and Putnam, T. (eds) *FutureNatural: Nature, Science, Culture*, Routledge, London

Williams, R. (1989) *The Science Show*, ABC Radio, 4 February

Wills, P. (1989) 'Genetic Information and the Determination of Functional Organisation in Biological Systems', *Systems Research*, 6, pp. 219–26

——(1994) 'Correcting Evolution: Biotechnology's Unfortunate Agenda', *Revue Internationale de Syste'mique*, 8, 4–5, pp. 455–68

——(1996) 'Transgenic Animals and Prion Diseases: Hypotheses, Risks, Regulations and Policies', *New Zealand Veterinary Journal*, 44, pp. 33–6

Wooster, R., Neuhausen S., Mangion J., Quirk Y., Ford D., Collins N., Nguyen, K. and Seal S. (1994) 'Localization of a breast cancer susceptibility gene, *BRCA2*, to chromosome 13q12–13', *Science*, 265, pp. 2088–90

Working Group on Sustainable Agriculture (1991) Sustainable Agriculture, CSIRO, Melbourne

Wright, S. (1986) 'Molecular Biology or Molecular Politics? The Production of Scientific Consensus on the Hazards of Recombinant-DNA Technology', *Social Studies of Science*, 16, pp. 593–620

——(1993) 'The Social Warp of Science: Writing the History of Genetic Engineering Policy', *Science, Technology and Human Values*, 18, 1, pp. 79–101

Wuensche, A. and Lesser, M. (1992) *The Global Dynamics of Cellular Automata*, Addison-Wesley, Reading, Massachusetts

Wynne, B. (1983) 'Redefining the Issues of Risk and Public Acceptance', *Futures*, February, pp. 13–32

——(1991) 'Knowledges in Context', *Science, Technology and Human Values*, 16, pp. 111–21

Yearley, S. (1996) *Sociology, Environmentalism, Globalization: Reinventing the Globe*, Sage, London

Zalucki, M: (1997) Associate Professor, Department of Entomology, University of Queensland, pers. comm., July

Zechauser, R. and Viscusi, W. (1996) 'The Risk Management Dilemma', *Annals of the American Academy of Political and Social Science*, 545, pp. 144–55

Zimmerman, L., Kendall, P., Stone, M. and Hoban, T. (1994) 'Consumer Knowledge and Concern about Biotechnology and Food Safety', *Food Technology*, 48, pp. 71–7

Zinnen, T. (1994) 'Risky Business: Issues in Teaching about Safety and Regulation', *Genetic Engineering News*, 14, 21, p. 32

Index